Placement and Routing
of Electronic Modules

T0203634

ELECTRICAL ENGINEERING-ELECTRONICS SOFTWARE

1. Transformer and Inductor Design Software for the IBM PC, *Colonel Wm. T. McLyman*
2. Transformer and Inductor Design Software for the Macintosh, *Colonel Wm. T. McLyman*
3. Digital Filter Design Software for the IBM PC, *Fred J. Taylor and Thanos Stouraitis*

Placement and Routing of Electronic Modules

edited by
Michael Pecht
CALCE Electronic Packaging Research Center
University of Maryland
College Park, Maryland

CRC Press
Taylor & Francis Group
Boca Raton London New York

CRC Press is an imprint of the
Taylor & Francis Group, an **informa** business

CRC Press
Taylor & Francis Group
6000 Broken Sound Parkway NW, Suite 300
Boca Raton, FL 33487-2742

First issued in paperback 2019

ISBN-13: 978-0-8247-8916-9 (hbk)
ISBN-13: 978-0-367-40242-6 (pbk)

Library of Congress Cataloging-in-Publication Data

Placement and routing of electronic modules / edited by Michael Pecht.
     p.   cm. -- (Electrical engineering and electronics ; 82)
    Includes bibliographical references and index.
    ISBN 0-8247-8916-4 (alk. paper)
    1. Printed circuits--Design and construction--Mathematical models.
    2. Electronic apparatus and appliances--Design and construction-
-Mathematical models. I. Pecht, Michael. II. Series.
TK7868.P7P57 1993
621.381'046--dc20   92-44796
                                                         CIP

To my parents, George and Dorothy Pecht,
my wife, Judy,
my children, Joann, Jefferson and Andrew,
and Grandma Shih.

Preface

High-density multilayer printed wiring boards and multi-chip modules are playing a critical role in the development of complex electronic systems. Since 1980, printed wire density has been skyrocketing, as multilayer techniques have emerged for high-density and high-speed applications. In the last decade, the wire density for multilayer substrates has jumped from two hundred wires/linear-inch to more than two thousand wires/linear-inch and the number of board layers has increased to more than sixty for high-speed computer applications. Routing electrical signals with a minimum of layers and vias is now challenging even the most complex combinatorial graph theories and the most sophisticated automation techniques. Furthermore, performance is no longer the only measure of success. In order to be competitive, reliability, ease of manufacture and cost effectiveness must also be incorporated into the placement and routing methods.

Unfortunately, for placement and routing, the figures of merit are often ill-defined, the goals are often ambiguous, and the methods are often inconsistent with the goals. The objectives of this book are to provide a scientific foundation for theories and techniques of placement and routing, to show how the theories compare, and to propose approaches for solving specific problems.

This book is for those who are technically serious about placement and routing of electronic devices and modules and capable of understanding and using the mathematics and theories. It is also for those who need practical methods and techniques to help place and route high-density modules.

The first of the book's eight chapters provides the mathematical concepts of set, combinatorial and graph theories for understanding the descriptions and theories in the following chapters. Because electronic signal nets are routed as a tree, the routing quality is tightly coupled to tree types; thus, Chapter 2 introduces the definitions, structures and relations of tree types and presents various methods of finding minimum trees for types often employed in placement and routing. Chapter 3 presents methods to estimate the workspace area needed to place and route the electronic modules. It also shows how to estimate the number of layers necessary to complete routing. Chapter 4 introduces a variety of placement methodologies for routability, as well as a theory for approximating the minimum Steiner tree. Chapter 5 focuses on placement techniques for reliability and manufacturability, with combined techniques for placement tradeoffs. Chapter 6 demonstrates a variety of search strategies for paths connecting two nodes on a workspace with obstacles. Chapter 7 discusses via minimization to reduce the workspace area, to ease the manufacture of the workspace, and to reduce the number of layers. Chapter 8 provides the theorems and techniques to address Steiner's problem with rectilinear distance.

Michael Pecht

Contents

Chapter 3: Signal Layer Estimation   59

Sudha Balakrishnan and Michael Pecht

Chapter 4: Placement for Routability   97

Yeun Tsun Wong, Michael Pecht, Michael D. Osterman and Guoqing Li

Chapter 5: Placement for Reliability and Producibility 139

Michael D. Osterman and Michael Pecht

Contributors

Guoqing Li received his Ph.D. in computer science from Clarkson University, majoring in complexity theory and software engineering. As a member of the CALCE EPRC, his research interests focus on system and software development for electronics design. He is a member of ACM and AMS.

Michael D. Osterman received his Ph.D. in mechanical engineering from the University of Maryland. He has worked with numerical techniques for placement, and optimization schemes for component placement based on reliability and routing constraints. Dr. Osterman is presently involved in the CALCE EPRC concurrent engineering effort. He is a member of ASME, ISHM, and IEEE.

Michael Pecht is a tenured faculty member at the University of Maryland, with a joint appointment in systems research and mechanical engineering. He is the chief editor of the *IEEE Transactions on Reliability*, sits on the editorial board of the *Journal of Electronics Manufacturing* and the *Journal of Concurrent Engineering*, and on the advisory board of the Society of Manufacturing Engineers, Electronics Division. Dr. Pecht is a Professional Engineer with a B.S. in acoustics, an M.S. in electrical engineering, and an M.S. and Ph.D. in engineering mechanics from the University of Wisconsin. He is an IEEE Fellow and the director of the CALCE EPRC.

Sudha Balakrishnan received an M.S. from the University of Maryland. Her research focuses on techniques and models for signal layer estimation.

Yeun Tsun Wong received an M.S. in mechanical engineering at the University of Maryland. His research focuses on tree theory and the applications in the algorithms for routing and placement. He and Dr. Pecht developed a theory to solve Steiner's problem with rectilinear distance.

About the CALCE EPRC

The Computer Aided Life Cycle Engineering (CALCE) Electronic Packaging Research Center (EPRC) at the University of Maryland is a consortium supported by industry and government members to conduct research and develop approaches for the design, simulation and assessment of reliable, cost-effective electronic products. The CALCE EPRC became a National Science Foundation (NSF) Industry-University Cooperative Research Center in July 1987 and a NSF State-Industry-University Cooperative Research Center in 1991.

The Center provides a unique world-class educational and research environment where students, faculty and industry personnel work together on problems of industrial importance in order to enhance industrial use of advanced techniques, to prepare students for industry, and to keep faculty and industry abreast and involved with current developments. The Center also uniquely provides industry with products (software and design guidelines) and services (technical assessments and training).

Placement and Routing of Electronic Modules

Chapter 1

Basic Concepts

Guoqing Li, Yeun Tsun Wong and Michael Pecht

Set theory, combinatorial theory and graph theory provide the neccessary mathematical foundation to place modules and route interconnects. These theories play an important role in developing algorithms to characterize electronic nets; expressing the relations between nets, the workspace, and placement constraints; and in formulating models to place modules and route electronic nets onto a workspace.

1.1 Sets

Set theory describes the basic relations among collections of objects, and many concepts in mathematics and computer science can be conveniently expressed in the language of sets. This section introduces the definitions, expressions, and properties of sets as they relate to placement and routing methods.

1.1.1 Notation

Suppose p and q are two arbitrary statements, and x and y are elements. The following notations indicate for the logical connectives in a mathematical proof.

(1) $p \wedge q$ means p *and* q. If $p \wedge q$ is true, both p and q must be true.

(2) $p \vee q$ means p *or* q. If p or q is true, then $p \vee q$ is true.

(3) $\bar{p}$ means *not* p. If p is true, then $\bar{p}$ is false and $\bar{\bar{p}}$ is true.

(4) $p \Rightarrow q$ means p *implies* q. That is, q must be true if p is true.

1

(5) $p \Leftrightarrow q$ means p and q are *equivalent*. That is, p and q are both true or both false: $p \Leftrightarrow q = (p \Rightarrow q) \wedge (q \Rightarrow p)$. For example, $(p \Rightarrow q) \Leftrightarrow (\overline{q} \Rightarrow \overline{p})$.

(6) $\exists x$ means *there exists* an x.

(7) $\forall x$ means *for all x* or *for every x*.

Example: For real numbers x and y, we define a proposition, p, by

$$p : (\forall x > 0)(\exists y) \quad (0 < y \wedge y < x)$$

which means that, for all positive real numbers x, there exists at least one positive number y such that $y < x$. Because we can find such a y for any given $x > 0$, proposition p is true.

◇

Example: Express the proposition below in words.

$$q : (\forall y < 0)(\exists x) \quad (x > 0 \wedge y \geq x).$$

This proposition states that for every negative y, there exists at least one positive number x which is equal to or less than y. Proposition q is false.

◇

1.1.2   Set definitions

A set is a collection of elements. If a set, A, consists of all x satisfying $p(x)$, then x is called an element of A. Mathematically, the relation between A and x is expressed as

$$A = \{x|\ p(x)\},$$
$$x \in A \ or \ A \ni x \tag{1.1}$$

where x is a member of A or x is contained by A. We use the notation $x \notin A$ or $A \not\ni x$ to mean x is not a member of A. If a set consists of elements $a, b, c, ...$, then the set can be expressed as

$$\{a, b, c, ...\} \tag{1.2}$$

The precedence of elements is ignored in a set. We call a set with a specified precedence of elements a *sequence*, and express it as

$$\{a, b, c, ...\}^* \tag{1.3}$$

Example: Let a set consist of three elements: 1, 2 and 3. Then
$$\{1,2,3\} = \{2,1,3\} = \{3,2,1\}$$
If a sequency consists of three elements: 1, 2 and 3, then
$$\{1,2,3\}^* \neq \{2,1,3\}^* \neq \{3,2,1\}^*$$
◇

An empty set, $\emptyset$, contains no elements and is expressed as:

$$\emptyset = \{x| \ x \neq x\} \tag{1.4}$$

Because no elements exist in an empty set, $x \notin A$ implies $x \notin \emptyset$.

Example: An empty set, $\emptyset$, can also be defined as:
$$\emptyset = \{x| \ x > a \ \wedge \ x < a\}$$
◇

1.1.3   Set algebra

If $x \in A$ implies $x \in B$, then A is defined as a subset of B. The relation between A and B is expressed as
$$A \subset B \ or \ B \supset A \tag{1.5}$$
Here, A equals B if and only if A and B contain the same elements. This statement can be expressed as $A = B \Leftrightarrow A \subset B \ \wedge \ B \subset A$. The following relations for sets are true.

$$\overline{\overline{A}} = A \tag{1.6}$$

$$A = B \ \Rightarrow \ B = A \tag{1.7}$$

$$A = B \ \wedge \ B = C \ \Rightarrow \ A = C \tag{1.8}$$

$$A \subset A \tag{1.9}$$

$$A \subset B \ \wedge \ B \subset C \ \Rightarrow \ A \subset C. \tag{1.10}$$

If $A \cup B = \{x \ | \ x \in A \ \vee \ x \in B\}$ and $A \cap B = \{x \ | \ x \in A \ \wedge \ x \in B\}$, then $A \cup B$ and $A \cap B$ are called the union and the intersection, respectively, of A and B. If $A \cap B \neq \emptyset$, then A and B intersect. From the definitions of union and intersection, the basic relations between A and B can be expressed:

$$A \ \subset \ A \cup B \tag{1.11}$$

$$A \cap B \subset A \tag{1.12}$$

$$A \cup B = A \Leftrightarrow B \subset A \tag{1.13}$$

$$A \cap B = A \Leftrightarrow A \subset B \tag{1.14}$$

Example: If $A = \{a, b, c, d\}$ and $B = \{c, d, e\}$, then $A \cup B = \{a, b, c, d, e\}$ and $A \cap B = \{c, d\}$.

◇

Common useful properties of union and intersection include the commutativity of sets:

$$\begin{cases} A \cup A = A \cap A = \emptyset \cup A \\ \emptyset \cap A = \emptyset \cap \overline{A} = \emptyset \end{cases} \tag{1.15}$$

$$\begin{cases} A \cup B = B \cup A \\ A \cap B = B \cap A \end{cases} \tag{1.16}$$

the associativity of sets:

$$\begin{cases} (A \cup B) \cup C = A \cup (B \cup C) \\ (A \cap B) \cap C = A \cap (B \cap C) \end{cases} \tag{1.17}$$

the distributivity of sets:

$$\begin{cases} A \cap (B \cup C) = (A \cap B) \cup (A \cap C) \\ A \cup (B \cap C) = (A \cup B) \cap (A \cup C) \end{cases} \tag{1.18}$$

and DeMorgan's Laws:

$$\begin{cases} \overline{A \cup B} = \overline{A} \cap \overline{B} \\ \overline{A \cap B} = \overline{A} \cup \overline{B}. \end{cases} \tag{1.19}$$

Example: This example shows an application of set algebra in the proof of DeMorgan's Law $\overline{A \cup B} = \overline{A} \cap \overline{B}$. The proof is given as follows.

If $x \in \overline{A \cup B}$, then $x \notin A \cup B$, and $x \notin A$ and $x \notin B$. Therefore $x \in \overline{A}$ and $x \in \overline{B}$, i.e., $x \in \overline{A} \cap \overline{B}$. Hence $\overline{A \cup B} \subset \overline{A} \cap \overline{B}$. Conversely, if $x \in \overline{A} \cap \overline{B}$, then $x \in \overline{A}$ and $x \in \overline{B}$. So $x \notin A$ and $x \notin B$. Therefore $x \notin A \cup B$, i.e., $x \in \overline{A \cup B}$. Hence $\overline{A} \cap \overline{B} \subset \overline{A \cup B}$. Figure 1.1 graphically depicts DeMorgan's Law:

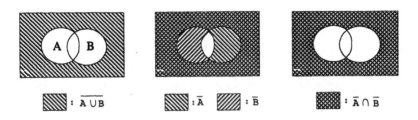

Figure 1.1. An example of DeMorgan's Law

◇

Another binary operation on sets A and B is set difference: $A - B = \{x |\, x \in A \wedge x \notin B\}$. This operation can be rewritten as $A - B = \{x |\, x \in A \wedge x \in \overline{B}\}$ and, finally, as

$$A - B = A \cap \overline{B} \tag{1.20}$$

Example: If $A = \{1,3,5,7\}$ and $B = \{1,5,8,9\}$, then $A - B = \{3,7\}$.

◇

The difference and the complement of sets have the following properties:

$$\begin{cases} A - \emptyset = A \\ A - A = \emptyset \end{cases} \tag{1.21}$$

$$\begin{cases} A - B \subset A \\ A \subset (A - B) \cup B \end{cases} \tag{1.22}$$

$$A \cap B = \emptyset \;\Rightarrow\; A - B = A \tag{1.23}$$

The total number of elements in the set $A \cup B$ equals the sum of the number of elements in set A and the number of elements in set B, minus the number of elements in $A \cap B$. Denoting the number of elements in set X as $|X|$,

$$|A \cup B| = |A| + |B| - |A \cap B|. \tag{1.24}$$

Example: If $A = \{1,3,5,7\}$ and $B = \{1,5,8,9\}$, then $A \cap B = \{1,5\}$ and $|A \cup B| = 4 + 4 - 2 = 6$.

◇

1.1.4 Boolean algebra

For binary systems, variables take only two values, 0 (false) and 1 (true). In Boolean algebra, the operators $\oplus$ and $\otimes$ are used instead of the $\cup$ and $\cap$ in set algebra. There are a total of sixteen operations in Boolean algebra. Common operations include *AND, OR, NOT, IMPLIES* and *EQUIVALENT*, which are denoted as $\cdot$, $+$, $-$, $\rightarrow$ and $\leftrightarrow$ respectively. The truth table of operations is charted in Table 1.1.

Table 1.1   Truth Table

p	q	$p \cdot q$	$p + q$	$\bar{p}$	$p \rightarrow q$	$p \leftrightarrow q$
0	0	0	0	1	1	1
0	1	0	1	1	1	0
1	0	0	1	0	0	0
1	1	1	1	0	1	1

From the truth table, the following useful propositions can be proven.

$$p \leftrightarrow q \Leftrightarrow (p \rightarrow q) \wedge (q \rightarrow p) \tag{1.25}$$

$$p \rightarrow q \Leftrightarrow \bar{p} \vee q \tag{1.26}$$

1.2   Combinatorial Mathematics

This section introduces the concepts of permutations and combinations [LIU68, CHR78], followed by two combinatorial problems which are related to routing. The first is associated with determining the number of rectilinear edges between two points with a specific number of bends. The second is concerned with finding the shortest tour for the traveling salesman problem. (For more background in this field, consult MIL72, KUN73, TRA76, COH78, GOU83.)

1.2.1   Permutations

If there are m ways to generate event E_1 and n ways to generate event E_2, then "event E_1 or event E_2" can be generated in $m + n$ ways and "event E_1 and event E_2" can be generated in $m \times n$ ways. These are called the *addition rule* and the *multiplication rule*, respectively.

Example: There are five odd numbers between 0 and 10. The number of sets containing one odd number is $n_1 = 5$. The number of sets containing two odd numbers is $n_2 = 10$. The total number of sets containing one or two odd numbers can be obtained using the addition rule: $n_1 + n_2 = 15$.

◇

Example: As shown in Figure 1.2, there are 3 distinct paths from p to q and 2 distinct paths from q to r. By the use of the multiplication rule, the number of paths from p to r through q is given as $n = 3 \times 2 = 6$.

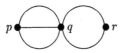

Figure 1.2. Paths from p to r through q

◇

Taking r elements from n elements in order is called the permutation of n distinct elements taken r at a time. The number of the permutations is denoted as $P(n, r)$. If one letter is taken away each time from a box that holds n distinct letters, then there are n different selections to take away the first letter, $n - 1$ selections to take away the second letter, ..., and $n - r + 1$ selections to take away the rth letter. Note that taking r letters away from the box is equivalent to taking the first letter away AND taking the second letter away AND ... AND taking the rth letter away. By the multiplication rule, the number of permutations of taking r letters away from n distinct letters is the number of selections for taking r letters away:

$$P(n,r) = n(n-1)(n-2)...(n-r+1) = \frac{n!}{(n-r)!} \qquad (1.27)$$

where $n! = n(n-1)(n-2)...3 \cdot 2 \cdot 1$. If $r = n$ then

$$p(n,n) = n! \qquad (1.28)$$

Example: Taking two letters from $\{A, B, C\}$ in order, the permutations are given in Table 1.2.

Table 1.2   An Example of Permutations

	A	B	C
A		AB	AC
B	BA		BC
C	CA	CB	

◇

1.2.2   Combinations

Taking r elements at a time from n distinct elements without considering their order is called the combination of n distinct elements taken r at a time. The number of combinations is denoted as $C(n, r)$. If we consider the order in the selections of r elements, the combination becomes a permutation. Therefore, the permutation of n distinct elements taken r elements at a time is equivalent to the combination of n distinct elements taken r elements at a time *AND* the permutation of r distinct elements taken r elements at a time. By the multiplication rule, the total number of permutations is given as

$$P(n, r) = r!C(n, r) \tag{1.29}$$

Therefore

$$C(n, r) = \frac{P(n, r)}{r!} = \frac{n!}{r!(n - r)!} \tag{1.30}$$

with

$$C(n, 0) = 1. \tag{1.31}$$

Example: Taking two letters at a time from $\{A, B, C\}$ without considering their order, the possible combinations are AB, AC and BC. These can be obtained from table 1.2 by setting $AB = BA$, $AC = CA$ and $BC = CB$.

◇

Equation 1.30 yields:

$$C(n, r) = C(n, n - r). \tag{1.32}$$

1.2.3   Rectilinear edges with a specific number of bends

Dividing a workspace into $(m + 1) \times (n + 1)$ square grids and locating point $p = (x_p, y_p)$ in the center of the lower left grid and point $q = (x_q, y_q)$ in the center of the upper right grid, the length of a rectilinear edge from p to q is the total length of $m + n$ grid edges, where $m = |x_p - x_q|$ and $n = |y_p - y_q|$ are the numbers of horizontal and vertical grid edges crossed by the rectilinear edge, respectively.

Example: Figure 1.3 shows a rectilinear edge from p to q crossing three horizontal grid edges and five vertical grid edges.

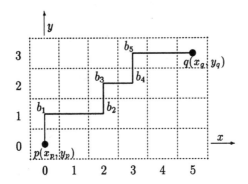

Figure 1.3. A rectilinear edge from p to q

◇

Let $b_i(x_i, y_i)$ be a bend point where a bend exists in the rectilinear edge. The coordinates x_i and y_i are grid indexes in x (horizontal) and y (vertical) directions, respectively. For example, $b_2(2, 1)$ is a bend point in Figure 1.3. The bend points are labeled in order, beginning with the one closest to p. Thus a rectilinear edge can be expressed as a sequence, comprised of point p, bend points and point q. For example, the rectilinear edge shown in Figure 1.3 can be expressed as a sequence $\{p, b_1, b_2, b_3, b_4, b_5, q\}^*$.

Assume a rectilinear edge from point p to q crosses m vertical grid edges and n horizontal grid edges. The grids are labeled from 0 to m and 0 to n in x and y directions, respectively. A rectilinear edge is determined by selecting m locations for vertical grid edges to be crossed or n locations for horizontal grid edges to be crossed from $m + n$ locations. *Determining a rectilinear edge from p to q is equivalent to determining m vertical grid edges from $m + n$ grid edges or determining n horizontal grid edges from $m + n$ grid edges.* Therefore, the number of distinct rectilinear edges between two points is the number of combinations of

$m + n$ elements taken m or n elements at a time. That is:

$$N_{edge} = \begin{cases} C(m+n,m) \\ C(m+n,n) \end{cases} \tag{1.33}$$

To find the maximum number of bends for a rectilinear edge from point p to q, the edge can be constructed as follows. If $m = n$, the edge can be constructed beginning at p and crossing a vertical grid edge first. The construction repetitively moves in sequences of *right* and *up* on the grids until reaching point q. Each sequence includes two grids where bends occur except the last one. The last sequence contains only one bend because the last move (*right*) in the sequence reaches point q. There are $2m$ moving sequences; thus, there are $2m - 1$ bends in the edge. The analysis is similar if the edge crosses a horizontal grid edge first. In this case, sequency will move *up* and *right* repetitively.

Example: Figure 1.4 shows how to find the maximum number of bends of a rectilinear edge if $m = n$. In the figures, $m = n = 3$ and the maximum number of bends is $2m - 1 = 5$.

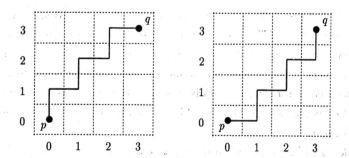

(a) Crossing a horizontal grid edge first. (b) Crossing a vertical grid edge first.

Figure 1.4. Finding maximum number of bends for $m = n = 3$

If $m \neq n$, without loss of generality we may assume $m > n$. The edge can be constructed in the same manner as above. The last moving sequence is not complete; includes only *moving right* and reaches the point q with no bend. The number of moving sequences (not including the last one) is bounded by n because $m > n$. Thus, there are $2n$ bends in the edge. Therefore, the maximum number of bends for a rectilinear edge from point p to q is

$$N_{maxbend} = \begin{cases} 2m - 1 & \text{if } m = n \\ 2min(m, n) & \text{if } m \neq n \end{cases} \tag{1.34}$$

Example: If $m = 5$ and $n = 3$, a rectilinear edge between p and q with the maximum number of bends can be drawn and $N_{maxbend} = 2 \times 3 = 6$. Note that this rectilinear edge is different from the one given in Figure 1.3 that contains five bends.

◇

The number of rectilinear edges from point p to point q with a specific number of bends can also be calculated. Let e_i be an edge with i bends, which crosses m vertical grid edges and n horizontal grid edges. Let N_{e_i} be the total number of e_i.

Case I. The number of bends, i, is even. Consider the edges that cross a vertical grid edge first. Bend points b_{2j-1} and b_{2j} have the same x-coordinates and b_{2j} and b_{2j+1} have the same y-coordinates for all $j > 1$. Given a set of x-coordinates of all odd-indexed bend points and a set of y-coordinates of all even-indexed bend points, all bend points are identified and hence the edge is determined. Because there are $\frac{i}{2}$ odd-indexed bend points and $m - 1$ positions (grids) that can be chosen as their x-coordinates, the number of ways to determine x-coordinates of all odd-indexed bend points is $C(m-1, \frac{i}{2})$. For the y-coordinates of the even-indexed bend points, although the number of the points is $\frac{i}{2}$, only $\frac{i}{2} - 1$ points need to be considered, because the last bend point has the same y-coordinate as point q. Because the number of candidates for the positions (grids) in the y direction is $n - 1$, the number of ways to determine the y-coordinates of all even-indexed bend points is $C(n-1, \frac{i}{2} - 1)$. Thus the number of edges with an even number of bends that cross a vertical grid edge first is $C(m-1, \frac{i}{2})C(n-1, \frac{i}{2} - 1)$.

Similarly, the number of the edges that cross a horizontal grid edge first is $C(n-1, \frac{i}{2})C(m-1, \frac{i}{2} - 1)$. Hence, the total number of the rectilinear edges from point p to q with an even number of bends is

$$C(m-1,\tfrac{i}{2})C(n-1,\tfrac{i}{2}-1) + C(n-1,\tfrac{i}{2})C(m-1,\tfrac{i}{2}-1) \qquad (1.35)$$

Example: An edge from point p to q with four bends is shown in Figure 1.5. It crosses a vertical grid edge first, with $m = 5$ and $n = 3$.

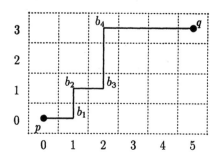

Figure 1.5. Illustration of Case I

◇

Case II. The number of bends, i, is odd. The analysis is similar to that for Case I. If the edges cross a vertical grid edge first, the x-coordinate of the last bend point is the same as that of point q. Thus the last odd-indexed bend point needn't be considered in determining the x-coordinates; $\frac{i-1}{2}$ odd-indexed bend points determine the x-coordinates. Because the number of possible grids for these x-coordinates is $m - 1$, the number of ways to determine the x-coordinates of the odd-indexed bend points is $C(m-1,\frac{i-1}{2})$.

There are $\frac{i-1}{2}$ bend points with even indexes. Because the number of possible grids for their y-coordinates is $n - 1$, there are $C(n-1,\frac{i-1}{2})$ ways to determine y-coordinates of even-indexed bend points. Thus the number of the edges with an odd number of bends that cross a vertical grid edge first is $C(m-1,\frac{i-1}{2})C(n-1,\frac{i-1}{2})$.

Example: An edge from point p to point q with five bends is shown in Figure 1.6. It crosses a vertical grid edge first, with $m = 5$ and $n = 3$.

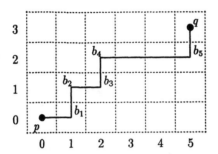

Figure 1.6. Illustration of Case II

◇

The result is the same for edges that cross a horizontal grid edge first; the number of such edges is $C(m-1, \frac{i-1}{2})C(n-1, \frac{i-1}{2})$. Hence the total number of rectilinear edges from point p to point q with an odd number of bends, i, is:

$$2C(m-1, \tfrac{i-1}{2})C(n-1, \tfrac{i-1}{2}) \tag{1.36}$$

Therefore, the number of the rectilinear edges that cross m vertical grid edges and n horizontal grid edges with i bends is:

$$N_{e_i} = \begin{cases} 2C(m-1, \frac{i-1}{2})C(n-1, \frac{i-1}{2}) & (i \text{ is odd}) \\ C(m-1, \frac{i}{2})C(n-1, \frac{i}{2}-1) + C(n-1, \frac{i}{2})C(m-1, \frac{i}{2}-1) & (i \text{ is even}) \end{cases} \tag{1.37}$$

Example: According to Equation 1.37, the total number of edges that crosses m vertical grid edges and n horizontal grid edges with five bends is $N_{e_5} = 2C(m-1,2)C(n-1,2)$. If $m = n = 3$, then $N_{e_5} = 2$. This means that there are two rectilinear edges with five bends that cross three vertical grid edges and three horizontal grid edges, as shown in Figure 1.4.

◇

1.2.4 The traveling salesman problem

Let G be a complete graph with n vertices representing connections between n cities, and let the set of vertices (cities) be $V = \{v_1, v_2, \ldots, v_n\}$. Every edge in G has a length representing the distance between two cities. $D = (d_{ij})_{n \times n}$ is the distance matrix of n cities, where d_{ij} is the distance from city v_i to city v_j (the length of the edge from vertex v_i to vertex v_j). A path starting at a given city, v_k, going through every city exactly once and finally returning to v_k will be called a *tour*. The length of a tour is the sum of the lengths of the edges on the path defining the tour. The traveling salesman problem is to find the tour of minimum length [AHO76, DU89].

Let $L_k(v_i; V - v_i)$ be the shortest length that begins at v_i, passes each vertex once in set $V - v_i$, and finally returns to v_k. The traveling salesman problem can be expressed as

$$L_k(v_k; V - v_k) = min(d_{kj} + L_k(v_j; V - v_k - v_j)) \quad (j = 1, 2, \ldots, n; j \neq k) \quad (1.38)$$

Example: Given a distance matrix:

$$
D = \begin{array}{c c}
 & \begin{array}{cccc} v_1 & v_2 & v_3 & v_4 \end{array} \\
\begin{array}{c} v_1 \\ v_2 \\ v_3 \\ v_4 \end{array} &
\left[\begin{array}{cccc}
0 & 8 & 5 & 6 \\
6 & 0 & 8 & 5 \\
7 & 9 & 0 & 5 \\
9 & 7 & 8 & 0
\end{array}\right]
\end{array},
$$

$L_1(v_1; v_2, v_3, v_4)$ means the length of the tour that begins at v_1, passes through v_2, v_3 and v_4 once, and finally returns to v_1. That is:

$$
\begin{aligned}
L_1(v_1; v_2, v_3, v_4) = \ & min(d_{12} + L_1(v_2; v_3, v_4), \\
& d_{13} + L_1(v_3; v_2, v_4), \\
& d_{14} + L_1(v_4; v_2, v_3)).
\end{aligned}
$$

Similarly, $L_1(v_2; v_3, v_4) = min(d_{23} + L_1(v_3; v_4), d_{24} + L_1(v_4; v_3))$. Corresponding to the given distance matrix:

$$
\begin{aligned}
L_1(v_2; v_1) &= d_{21} = 6 \\
L_1(v_3; v_1) &= d_{31} = 7 \\
L_1(v_4; v_1) &= d_{41} = 9 \\
L_1(v_2; v_3) &= d_{23} + L_1(v_3; v_1) = 8 + 7 = 15 \\
L_1(v_2; v_4) &= d_{24} + L_1(v_4; v_1) = 5 + 9 = 14 \\
L_1(v_3; v_2) &= d_{32} + L_1(v_2; v_1) = 9 + 6 = 15
\end{aligned}
$$

$$L_1(v_3; v_4) = d_{34} + L_1(v_4; v_1) = 5 + 9 = 14$$
$$L_1(v_4; v_2) = d_{42} + L_1(v_2; v_1) = 7 + 6 = 13$$
$$L_1(v_4; v_3) = d_{43} + L_1(v_3; v_1) = 8 + 7 = 15$$

$$L_1(v_2; v_3, v_4) = min(d_{23} + L_1(v_3; v_4), d_{24} + L_1(v_4; v_3))$$
$$= min(8 + 14,\ 5 + 15) = 20$$
$$L_1(v_3; v_2, v_4) = min(d_{32} + L_1(v_2; v_4), d_{34} + L_1(v_4; v_2))$$
$$= min(9 + 14,\ 5 + 13) = 18$$
$$L_1(v_4; v_2, v_3) = min(d_{42} + L_1(v_2; v_3), d_{43} + L_1(v_3; v_2))$$
$$= min(7 + 15,\ 8 + 15) = 22$$

Finally,

$$L_1(v_1; v_2, v_3, v_4) = min(d_{12} + L_1(v_2; v_3, v_4), d_{13}$$
$$+ L_1(v_3; v_2, v_4)), d_{14} + L_1(v_4; v_2, v_3))$$
$$= min(8 + 20,\ 5 + 18,\ 6 + 22)$$
$$= d_{13} + L_1(v_3; v_2, v_4)$$
$$= 23$$

Therefore, the shortest tour beginning at v_1 is $v_1 \rightarrow v_3 \rightarrow v_4 \rightarrow v_2 \rightarrow v_1$. The total length of the tour is 23.

◇

Equation 1.38 is a recursive formula. The computation process in the above example can be expressed as a search tree (Figure 1.7).

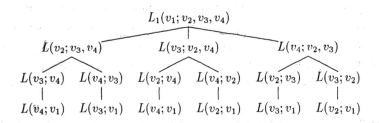

Figure 1.7. Search tree for the recursive process

The computation of each parent node is based on the computation of its son nodes in the search tree. At the first level (top level), $n-1$ addition operations are required to compute $L_1(v_1; V - v_1)$. At the second level, the number of $L_1(v_{i_1}; V - \{v_1, v_{i_1}\})$ $(i_1 = 2, 3, \ldots, n)$ is $(n-1)C(n-2, n-2)$ and each requires $n-2$ addition operations. At the third level, the number of $L_1(v_{i_2}; V - \{v_1, v_{i_1}, v_{i_2}\})$ $(i_2 = 2, 3, \ldots, n; i_2 \neq i_1)$ is $(n-1)C(n-2, n-3)$ and each requires $n-3$ addition operations. At the kth level, the number of $L_1(v_{i_{k-1}}; V - \{v_1, v_{i_1}, v_{i_2}, \ldots, v_{i_{k-1}}\})$ is $(n-1)C(n-2, k-1)$ and each requires $n-k$ addition operations. Therefore, the total number of addition operations is:

$$
\begin{aligned}
M \;=\; & (n-1) + (n-1)[(n-2)C(n-2, n-2) \\
& +(n-3)C(n-2, n-3) \\
& +\ldots + C(n-2, 1)]
\end{aligned}
$$

Further simplification gives:

$$
M = (n-1) + (n-1)(n-2)2^{n-3} \tag{1.39}
$$

Thus, the time complexity to find a tour of minimum length is exponential to the number of cities.

1.2.5   The NP theory

The languages recognizable in deterministic polynomial time form a natural and important class, the class $\bigcup_{i \geq 1} \mathrm{DTIME}(n^i)$, which is denoted by $\mathcal{P}$. It is an intuitively appealing notion that $\mathcal{P}$ is a class of problems that can be solved efficiently. A number of important problems do not appear to be in $\mathcal{P}$ but have efficient nondeterministic algorithms. These problems fall into the class $\bigcup_{i \geq 1} \mathrm{NTIME}(n^i)$, which is denoted by $\mathcal{NP}$. For example, the traveling salesman problem is an NP problem. All known algorithms for solving the traveling salesman problem require, like the one discussed in the previous subsection, an exponential amount of time. The theory for NP problems is developed to prove the equivalence of the time complexity of different problems, rather than to find algorithms to solve the problems.

If there exists a function that can complete the transformation from problem P_1 to problem P_2 in polynomial time, then it is said that P_1 can be transformed to P_2 in polynomial time, and the transformation is notated as $P_1 \propto P_2$. If x is an input and $f(x)$ is a function that can complete the transformation from P_1 to P_2 in polynomial time, then $(P_1 \propto P_2) \Leftrightarrow (x \in P_1 \Rightarrow f(x) \in P_2)$.

Lemma 1.1 If $P_1 \propto P_2$ and $P_2 \in \mathcal{P}$, then $P_1 \in \mathcal{P}$.

Proof: Because $(P_1 \propto P_2) \Leftrightarrow (x \in P_1 \Rightarrow f(x) \in P_2)$ and $f(x)$ has a polynomial time bound, the output from $f(x)$ must also have a polynomial bound, denoted as g. Assume the polynomial time bound of P_2 is h. The process from the input of P_1 to the output of P_2 can be expressed as

The total time from the input of P_1 to the output of P_2 is $g(|x|) + h(g(|x|))$ which has a polynomial bound. Therefore, $P_1 \in \mathcal{P}$. $\square$

Polynomial time transformations combine in transitive fashion. That is, if $P_1 \propto P_2$ and $P_2 \propto P_3$, then $P_1 \propto P_3$. A problem, P_1, is called NP-complete if and only if:

1) $P_1 \in \mathcal{NP}$.
2) For every problem, $P' \in \mathcal{NP}$, there is a polynomial time transformation from P' to P_1.

Let $A_1, A_2, ...$, denote Boolean variables (either true or false). Let $\overline{A_1}, \overline{A_2}, ...$, denote the negation of $A_1, A_2, ...$, respectively. A *literal* is either a variable or its negation. A formula in the propositional calculus is an expression that can be constructed using literals and the logical operations *AND* and *OR*. An arbitrary logical expression can be expressed in the conjunctive normal form by using Boolean algebra (section 1.1.4). The satisfiability problem (SAT) is to determine if a formula is true for some assignment of truth values to the variables.

Example: Transform the logical expression $\overline{F} \vee (\overline{B} \vee C) \vee D$ into the conjunctive normal form.

$$\overline{F} \vee \overline{(\overline{B} \vee C)} \vee D$$
$$= (\overline{F} \vee (B \wedge \overline{C})) \vee D$$
$$= ((\overline{F} \vee B) \wedge (\overline{F} \vee \overline{C})) \vee D$$
$$= D \vee ((\overline{F} \vee B) \wedge (\overline{F} \vee \overline{C}))$$
$$= (D \vee (\overline{F} \vee B)) \wedge (D \vee (\overline{F} \vee \overline{C}))$$
$$= (D \vee \overline{F} \vee B) \wedge (D \vee \overline{F} \vee \overline{C})$$

◇

In 1972, S. Cook developed a theory to prove the relationship between NP problems and satisfiability problem. Cook's theorem can be proved by Lemma 1.1 [GAR79].

Theorem 1.1 (Cook's Theorem) If $L \in \mathcal{NP}$, then $L \propto SAT$.
Corollary 1.1.1 $\mathcal{P} = \mathcal{NP}$ if and only if $SAT \in \mathcal{P}$.

If there exists a solution with a polynomial time bound for the satisfiability problem, then all NP problems become P problems because all NP problems can be transformed to the satisfiability problem. The satisfiability problem is also called an NP-complete problem — historically, the first such problem. To prove a problem, P_1, to be NP-complete, the following two steps are usually required:

1) Prove problem P_1 is an NP problem.
2) Prove that another problem, P_2, which is an NP-complete problem, can be transformed to problem P_1 in polynomial time.

From Cook's theorem, if one NP-complete problem is proved to have a polynomial time algorithm, then all the NP problems have polynomial time algorithms. Thus the existence of a polynomial time algorithm for just one NP-complete problem would imply $\mathcal{P} = \mathcal{NP}$. Unfortunately, so far it cannot be proven whether or not a polynomial time algorithm exists for an NP-complete problem. Therefore, no NP-complete problem can be solved by any known polynomial time algorithm.

For every problem, P', in $\mathcal{NP}$, if $P' \propto \pi$, then π is called an NP-hard problem. An NP-hard problem can be transformed from an NP-complete problem. However, an NP-hard problem does not have to be an NP-complete problem. An optimization problem that corresponds to an NP-complete decision problem is usually an NP-hard problem.

1.3 Graph Theory

Graph theory describes the relations between vertices and their interconnections [ALA88]. A printed wire board can be treated as a graph system in which terminals (pins) of modules are considered to be vertices and wires can be interconnections between terminals of modules. A *circuit* on the board can be defined by a set of modules *M*, a set of terminals *T*, and a set of nets *N*. A *terminal* is either an input or output pin on the boundaries of the chips or modules. A *net* (signal net) is a set of terminals to be interconnected by conductive paths (wires). A *wire segment* is a line segment on a specified layer which implements all or part of a net. A point, other than a terminal, at which two or more wire segments meet and are electrically connected is called a *junction*. The junction degree is the number of wire segments joined at a particular junction. This section presents the basic definitions and concepts of the graph theory that will be employed in the discussions of trees, placement, and routing methods. (For elementary graph theory, consult [BOR85, HAR73, HAR69]).

1.3.1 Graphs

A graph is defined to be a pair (V, E), where $V = \{v_1, v_2, ..., v_p\} \neq \emptyset$ $(p = |V|)$ and $E = \{e_1, e_2, ..., e_q\}$ $(q = |E|)$, in which v_i $(i = 1, 2, ..., p)$ is a vertex (point or node) and e_i $(i = 1, 2, ..., q)$ is an edge (a line or a curve). $G = (V, E)$ is also called a $(|V|, |E|)$ graph. The edge $e = uv \in E$ $(u \neq v)$ is said to join the adjacent vertices $u \in V$ and $v \in V$. Note that $uv = vu$. If $e_1 = uv_1$ and $e_2 = uv_2$ $(v_1 \neq v_2)$, then e_1 and e_2 are adjacent edges. If each bend in edge uv is a right angle, uv is particularly denoted as e_{uv}.

Example: Figure 1.8 shows a graph, $G = (V, E)$, in which $V = \{v_1, v_2, ..., v_5\}$ and $E = \{v_1v_2, v_1v_4, v_1v_5, v_2v_3, v_3v_5\}$. Note that the bend in edge v_2v_3 ($e_{v_2v_3}$) is a right angle, and distance measures for edges v_1v_2, v_1v_5 and v_2v_3 are different.

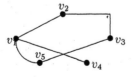

Figure 1.8. An example of a graph

◇

To distinguish different graphs, $V(G)$ and $E(G)$ are used to represent the set of vertices and the set of edges, respectively, in G. Different geometric diagrams often represent the same graph using the concept of isomorphism. If there exists a one-to-one mapping between $V(G_1)$ and $V(G_2)$ that preserves the adjacency and nonadjacency of vertices in $V(G_1)$ and $V(G_2)$, then graphs G_1 and G_2 are said to be isomorphic, and are expressed as $G_1 \cong G_2$. For a graph G, if $V(G)$ can be partitioned into two sets V_1 and V_2 such that no two vertices in the same set are adjacent, then $G = (V_1, V_2, E)$ is a bipartite graph.

Example: Graphs G_1 and G_2 in Figure 1.9 are isomorphic because there is a one-to-one mapping between $V(G_1)$ and $V(G_2)$:

$$v_1 \Leftrightarrow u_1,\ v_2 \Leftrightarrow u_3,\ v_3 \Leftrightarrow u_5,\ v_4 \Leftrightarrow u_2,\ v_5 \Leftrightarrow u_4,\ v_6 \Leftrightarrow u_6$$

The relationships in G_1 are preserved in G_2 as follows:

$$v_1v_4 \Leftrightarrow u_1u_2,\ v_1v_5 \Leftrightarrow u_1u_4,\ v_1v_6 \Leftrightarrow u_1u_6,\ v_2v_4 \Leftrightarrow u_3u_2,\ v_2v_5 \Leftrightarrow u_3u_4,\ v_2v_6 \Leftrightarrow u_3u_6,$$
$$v_3v_4 \Leftrightarrow u_5u_2,\ v_3v_5 \Leftrightarrow u_5u_4,\ v_3v_6 \Leftrightarrow u_5u_6$$

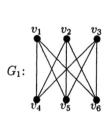 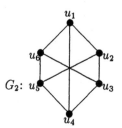

Figure 1.9. Isomorphic graphs

◇

Example: Graphs G_1 and G_2 in Figure 1.9 are bipartite graphs, in which $V_1(G_1) = \{v_1, v_2, v_3\}$, $V_2(G_1) = \{v_4, v_5, v_6\}$, $V_1(G_2) = \{u_1, u_3, u_5\}$, and $V_2(G_2) = \{u_2, u_4, u_6\}$.

◇

The complement, $\overline{G}$, of a graph, G, is the graph with the same vertex set in which two vertices are adjacent in $\overline{G}$ if and only if these vertices are not adjacent in G. $G \cup \overline{G}$ is called a complete graph, in which any two of its vertices are adjacent or any vertex is adjacent to all other vertices of G. If G is a (p,q) graph, then $\overline{G}$ is a $(p,\overline{q})$ graph and $G \cup \overline{G}$ is denoted as K_p, where $q + \overline{q} = p(p-1)/2$, because each vertex has $p-1$ incident edges and all edges are repeated twice in a complete graph. If each $v_1 \in V_1$ is adjacent to all vertices in V_2 and vice versa in $G = (V_1, V_2, E)$, then G is called a complete bipartite graph and is denoted as $K_{m,n}$ ($|V_1| = m$ and $|V_2| = n$). A graph G is self-complementary if $G \cong \overline{G}$.

Example: Figure 1.10 shows self-complementary graphs and the corresponding complete graph.

Figure 1.10. Self-complementary and complete graphs

◇

If $V(H) \subseteq V(G)$ and $E(H) \subseteq E(G)$, then graph H is a subgraph of graph G and G is a supergraph of H. Any graph isomorphic to a subgraph of G is also referred to as a subgraph of G. The simplest type of subgraph of a graph, G, is obtained by deleting a vertex or an edge from G. If $v \in V(G)$ and $|V(G)| \geq 2$, then $G - v$ denotes the subgraph with vertex set $V(G) - \{v\}$ and all edges that are not incident with v in G. If $e \in E$, then $G - e$ is the subgraph having $V(G)$ and $E(G) - \{e\}$.

Example: Figure 1.11 shows a supergraph G and its subgraphs.

Figure 1.11. A supergraph and its subgraphs

◇

1.3.2   The degree of a vertex

The degree or valency of a vertex, v, in a graph, G, is the number of edges incident with v and is represented as $deg\ v$. For instance, $deg\ v = 2$ in Figure 1.11. If $deg\ v = 0$, v is an isolated vertex. If $deg\ v = 1$, v is an end vertex. Because every edge is incident with two vertices and each edge is counted twice, the sum of the degrees of all the vertices is

$$\sum_{i=1}^{p} deg\ v_i = 2q \qquad (1.40)$$

where $V(G) = \{v_1, v_2, ..., v_p\}$ and $E(G) = \{e_1, e_2, ..., e_q\}$. Based on Equation 1.40, it can be proven that the number of vertices with odd degrees is even.

An r-regular graph, G, is regular of degree r if $deg\ v = r$ for every vertex, v, in G. A complete (p, q) graph is a regular graph of degree $p - 1$. Every graph is a subgraph of a regular graph.

Example: Regular graphs with degrees from 0 to 3 are shown in Figure 1.12.

$G_0:$ $G_1:$ $G_2:$ $G_3:$

Figure 1.12. Regular graphs

◇

1.3.3   Line graphs

For $G' = (V(G'), E(G'))$, if $v \in V(G')$ corresponds to edge $e \in E(G)$, v is said to be a correspondent vertex of e. If a one-to-one correspondence exists between $E(G)$ and $V(G')$, and if the adjacency of each two vertices of $V(G')$ corresponds to the adjacency of their corresponding edges of $E(G)$, then G' is called a line graph or edge graph of G, and is denoted as $L(G)$ (Figure 1.13). If $e = uv \in E(G)$, then $deg\ e$ in $L(G)$ is $deg\ u + deg\ v - 2$.

Example: Figure 1.13 shows line graphs, where $G_2 = L(G_1)$ and $L(G_2) = L(L(G_1)) = L^2(G_1)$.

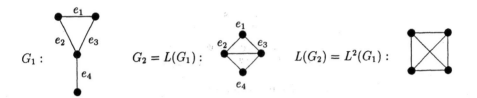

Figure 1.13. Line graphs

◇

There are q vertices in $L(G)$ for a (p,q) graph G. Because $deg\ e_i$ edges that are incident with vertex $e_i \in G$ will produce $C(deg\ e_i, 2)$ edges in $L(G)$, the total number of edges in $L(G)$, q_L, is

$$
\begin{aligned}
q_L &= \sum_{i=1}^{p} C(deg\ e_i, 2) \\
&= \frac{1}{2}\sum_{i=1}^{p}[(deg\ e_i)^2 - deg\ e_i] \\
&= \frac{1}{2}\sum_{i=1}^{p}(deg\ e_i)^2 - q
\end{aligned}
\tag{1.41}
$$

Theorem 1.2 If G_1 and G_2 are two nontrivial connected graphs which are not K_3 and $K_{1,3}$, then $L(G_1) \cong L(G_2)$, if and only if $G_1 \cong G_2$. (proof omitted)

1.3.4 Paths, cycles and trees

A finite alternative sequence of vertices and edges of G, beginning at vertex u and ending with vertex v, is called a u-v walk. A walk with no edges is a trivial walk. If $u = v$ and no vertex is repeated between u and v, the non-trivial walk u-v is called a cycle. If $u \neq v$ and no vertex is repeated, the walk is called a path. If $u \neq v$ and no edge is repeated, the walk is called a trail. The length of a path (cycle) is its number of edges.

Example: As shown in Figure 1.14, $\{v_1, v_2, v_5, v_4, v_2, v_3\}^*$ is a trail, $\{v_1, v_2, v_5, v_4\}^*$ is a path and $\{v_2, v_4, v_5, v_2\}^*$ is a cycle. Note that $\{v_1, v_2, v_4, v_5, v_2\}^*$ is not a cycle because $v_1 \neq v_2$.

Figure 1.14. Walks

◊

Two vertices, u and v, of a graph, G, are said to be connected if there exists a u-v path in G. Graph G is connected if every two of its vertices are connected. If a connected graph, G, is separated into two or more disconnected subgraphs by removing a vertex, v, from G, then v is called a cut vertex. A vertex, v, in a connected graph G is a cut vertex if and only if there exist vertices u and w ($u, w \neq v$) such that v is on every u-w path of G.

If a connected graph, G, is separated into two disconnected subgraphs by removing an edge, e, from G, then e is called a bridge. An edge, e, in a connected graph, G, is a bridge if and only if there exist vertices u and w such that e is on every u-w path of G. An edge of a cycle cannot be a bridge and an edge that does not belong to a cycle must be a bridge. A nontrivial connected graph with no cut vertices is called a block.

Example: In Figure 1.14, v_2 is a cut vertex; edges v_1v_2 and v_2v_3 are bridges, but v_2v_4, v_2v_5 and v_4v_5 are not.
◊

Example: A graph G and its blocks are shown in Figure 1.15.

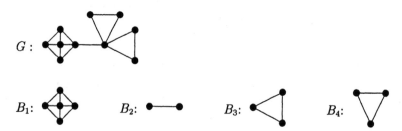

Figure 1.15. Graph G and its blocks
◊

If e is an edge of a connected graph, G, the following statements are equivalent:

(a) $e \in G$ is a bridge.

(b) e does not exist in any cycle of G.

(c) e exists in every path that connects vertices $u \in G$ and $v \in G$.

(d) There is a partition that divides $V(G)$ into subsets U and W such that e exists in every path that connects vertices $u \in U$ and $w \in W$.

If v is a vertex of a connected graph, G, the following statements are equivalent:

(a) $v \in G$ is a cut vertex.

(b) v exists in every path that connects vertices $u \in G$ and $w \in G$.

(c) There is a partition that divides $V(G) - \{v\}$ into subsets U and W such that v exists in every path that connects vertices $u \in U$ and $w \in W$.

If there are at least three vertices in a connected graph, G, the following statements are equivalent:

(a) G is a block.

(b) Any two vertices of G exist in the same cycle.

(c) Any vertex and any edge of G exist in the same cycle.

(d) Any two edges of G exist in the same cycle.

(e) There exists a path that connects two given distinct vertices and passes a given edge.

(f) There exists a path that connects two given distinct vertices and passes the third given vertex.

(g) There exists a path that connects two given distinct vertices and does not pass the third given vertex.

A tree is a graph without cycles. Let $G = (V, E)$ be a $(|V|, |E|)$ graph with $|V| \geq 2$. The following statements are then equivalent and each characterizes a tree:

(a) G is connected and has no cycles.

(b) Only one path exists between $u \in V(G)$ and $v \in V(G)$.

(c) G is connected and $p = q + 1$.

(d) G has no cycle and $p = q + 1$.

(e) G has no cycle. There is a cycle in $G + uv$ for any u and $v \in V(G)$ that are not adjacent.

1.3.5   The adjacency matrix

To represent connections between vertices and to manipulate them within a computer, a labeled graph can be represented by the adjacency matrix. The adjacency matrix, A, of graph G with a set of vertices $V = \{v_1, v_2, ..., v_p\}$ is a $p \times p$ matrix $[a_{ij}]$, where $a_{ij} = 1$ if $v_iv_j \in E(G)$, and $a_{i,j} = 0$ if $v_iv_j \notin E(G)$.

Example: The graph shown in Figure 1.14 can be expressed as the following adjacent matrix:

$$A = \begin{bmatrix} 0 & 1 & 0 & 0 & 0 \\ 1 & 0 & 1 & 1 & 1 \\ 0 & 1 & 0 & 0 & 0 \\ 0 & 1 & 0 & 0 & 1 \\ 0 & 1 & 0 & 1 & 0 \end{bmatrix}.$$

◇

If $a_{ij} = a_{jk} = a_{ki} = 1$ for columns and rows i, j and k of A, then a cycle exists in graph G. For example, $a_{24} = a_{45} = a_{52} = 1$, and therefore, a cycle $\{v_2, v_4, v_5, v_2\}^*$ exists (Figure 1.14). Graph G is disconnected if and only if G can be represented by a matrix in the form

$$A = \begin{bmatrix} A_{11} & 0 \\ 0 & A_{22} \end{bmatrix}$$

where A_{11} and A_{22} are square matrices.

Example: If v_2v_4 and v_2v_5 are removed from the graph shown in Figure 1.14, then

$$A_{11} = \begin{bmatrix} 0 & 1 & 0 \\ 1 & 0 & 1 \\ 0 & 1 & 0 \end{bmatrix}$$

and

$$A_{22} = \begin{bmatrix} 0 & 1 \\ 1 & 0 \end{bmatrix}$$

where A_{11} and A_{22} are two matrices of subgraphs G_1 and $G_2 \in G$, in which $V(G_1) = \{v_1, v_2, v_3\}$, $E(G_1) = \{v_1v_2, v_2v_3\}$, $V(G_2) = \{v_4, v_5\}$ and $E(G_2) = v_4v_5$. Because $V(G_1)$ and $V(G_2)$ cannot be connected by any edge, G_1 is disconnected from G_2.

◇

1.4 References

[AHO76] Aho, A. V., Hopcroft, J. E., and Ullman, J. D., *The Design and Analysis of Computer Algorithms*, Addison-Wesley, Reading, Massachusetts, 1976.

[ALA88] Alavi, Y., Chartrand, G., Qellerman, O. R., and Schwenk, A. J., *Graph Theory, Combinatorics, and Applications*, John Wiley & Sons, New York, Vol. 1 and Vol. 2, 1988.

[BOR85] Berge, C., *Graphs*, North-Holland, New York, 1985.

[CHR78] Christofides, N., *Combinatorial Optimization*, John Wiley & Sons, New York, 1978.

[COH78] Cohen, D. I. A., *Basic Techniques of Combinatorial Theory*, John Wiley & Sons, New York, 1978.

[DU89] Du, D. Z., and Hu, G. D., *Combinatorics, Computing and Complexity*, Science Press, Beijing, Kluwer Academic Publishers, Boston, 1989.

[GAR79] Garey, M. R., and Johnson, D. S., *Computers and Intractability, A Guide to the Theory of NP-Completeness*, Freeman, San Francisco, 1979.

[GOU83] Goulden, I. P., and Jackson, D. M., *Combinatorial Enumeration*, John Wiley & Sons, New York, 1983.

[HAR69] Harray, H., *Graph Theory*, Addison-Wesley, Reading, Massachusetts, 1969.

[HAR73] Harray, H., and Palmer, E. M., *Graphical Enumeration*, Academic Press, New York, 1973.

[KUN73] Knuth, D. E., *The Art of Computer Programming*, Vol. 3, Addison-Wesley, Reading, Massachusetts, 1973.

[LIU68] Liu, C. L., *Introduction to Combinatorial Mathematics*, McGraw-Hill, New York, 1968.

[MIL72]   Miller, R. E., and Thatcher, J. W., *Complexity of Computer Computations*, Plenum Press, New York, 1972.

[TRA76]   Traub, J. F., *Algorithms and Complexity*, Academic Press, New York, 1976.

Chapter 2

Characterization and Generation of Trees

Yeun Tsun Wong, Guoqing Li and Michael Pecht

The goal of placement and routing is to provide the most effective connection of signals among devices, whether they are transistors on a chip or integrated circuits or modules on a printed circuit or wiring board. For these planar architectures, the approach to the connection of signals will determine the effectiveness of the placement and routing algorithms. The basis for the connection architecture is called the tree type, and a connection of signals is often required to be routed as a tree with the minimum length.

2.1 Tree Types

The terminal locations of a signal set represent nodes or vertices in graph theory. An edge can be formed between any two known nodes or extra vertices. In a tree, edges are not allowed to form any cycles. Based on different length measures of an edge, trees can be categorized as Euclidean and rectilinear (Manhattan) trees within a tree type. Tree types employed in the interconnection schemes include the Steiner tree, the spanning tree, the chain tree, the source-sink tree, the row-based tree and special trees.

2.1.1   Routing length

The most common interconnection norm is the total wire or routing length. The common distance function used to measure routing length is the rectilinear or Manhattan distance. For two connections located at positions (x_1, y_1) and (x_2, y_2), the rectilinear distance is

$$d_{12} = |x_1 - x_2| + |y_1 - y_2| \qquad (2.1)$$

Another distance function that can be used is the Euclidean distance:

$$D_{12} = \sqrt{(x_1 - x_2)^2 + (y_1 - y_2)^2}. \qquad (2.2)$$

Here, $2\frac{1}{2}$-dimensional is considered, because the out-of-plane dimension is usually perpendicular to the plane, and has smaller dimensions then that in-plane routing space. For a planar workspace, such as a board or substrate, the out-of-plane wiring distance, $z_1 - z_2$, is nearly zero and is therefore neglected, even if the board is multilayered.

An edge connecting two nodes, p and q, with a rectilinear distance is denoted e_{pq}. If the length of every edge in a tree is measured by the rectilinear distance, then the tree is called a rectilinear tree. As discussed above (Equation 1.33), the total number of shortest rectilinear edges connecting (x_1, y_1) and (x_2, y_2) is $C(|x_1 - x_2| + |y_1 - y_2|, |x_1 - x_2|)$. In this book, the rectilinear, rather than the Euclidean distance is emphasized, because electrical nets are almost always printed as rectilinear trees.

Example: The Euclidean distance measure is used in Figure 2.1a, and the rectilinear distance measure is used in Figure 2.1b. Note that, as expected, there is a reduction in routing length with the Euclidean distance measure. However, routing with Euclidean distance would cause severe over-talking, and the number of layers needed for isolating the over-talks cannot be tolerated.

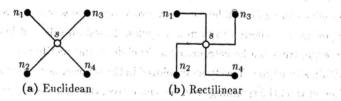

(a) Euclidean   (b) Rectilinear

Figure 2.1. Euclidean and rectilinear distances

2.1.2 Steiner trees

In the early nineteenth century, Jacob Steiner, the famous representative of geometry at the University of Berlin, added extra vertices to the spanning tree to reduce the tree length [COU41]. The extra vertex is called a Steiner point, and a tree with Steiner points is called a Steiner tree. A Steiner tree, connecting $V(t) = V_n \cup V_s$ with a set of edges $E(t)$, can be expressed as $t = (V(t), E(t))$, where $V_n = \{n_1, n_2, \ldots, n_{n_n}\}$ is a set of nodes, and $V_s = \{s_1, s_2, \ldots, s_{n_s}\}$ is a set of Steiner points. The length of tree t is denoted as l_t.

A Steiner tree with the minimum length is called a minimum Steiner tree. A number of minimum Steiner trees may be generated from the same node configuration. Finding minimum Steiner trees is called Steiner's problem [COU41, MEL61, HAN66, GIL68, COC72, DRE72, HAK72, HAW76, GAR77]. If every edge is rectilinear, the minimum Steiner tree is called a minimum rectilinear Steiner tree (MRST). The routing with a minimum of layers is dominated by Steiner's problem with rectilinear distance [KUH86]. Figure 2.2 shows two Steiner trees for the same node configuration. Figure 2.2b depicts a minimum Steiner tree.

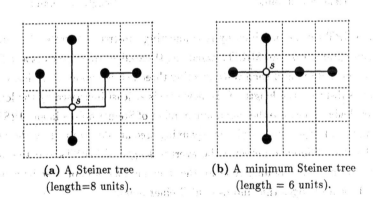

(a) A Steiner tree
(length=8 units).

(b) A minimum Steiner tree
(length = 6 units).

Figure 2.2. Steiner trees

2.1.3 Spanning trees

The spanning tree is a special case of the Steiner tree, where no Steiner points exist ($V_s = \emptyset$). Examples of a spanning tree for the configuration of nodes given in Figure 2.2 are depicted in Figures 2.3 and 2.4. A spanning tree with a minimum length is called a minimum spanning tree (MSPT). Figure 2.4 is an MSPT. The minimum spanning tree length has typically been employed as a measure of a placement procedure's success.

The length of a minimum Steiner tree is equal to or shorter than the length of a minimum spanning tree for the same node configuration. For the same node configuration, there are also more alternative Steiner trees with the MRST length then the alternative MSPTs.

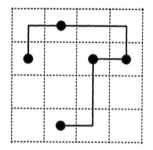

Figure 2.3. A spanning tree (length = 9 units).

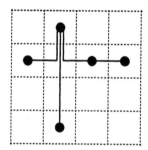

Figure 2.4. A minimum spanning tree (length = 8 units)

An MSPT can be obtained by re-connecting sub-trees generated by removing all Steiner points from an MRST. Removing three edges that intersect at a Steiner point in an MRST, and then re-connecting three nearest nodes (terminals) of the separate sub-trees, the length of the new tree is at least increased by the length of a removal edge. Because the maximum number of Steiner points in an MRST, and the number of edges in a minimum spanning tree are determined by the number of nodes, the difference between the average length of MRSTs and the average length of minimum spanning trees for the same number of nodes can be estimated by the length of edges that intersect at Steiner points.

Example: Figure 2.5a depicts an MRST for three nodes. Figure 2.5b depicts a minimum spanning tree for the same node configuration, that is generated by removing $e_{n_1 s}$, $e_{n_2 s}$ and $e_{n_3 s}$ from the MRST, and reconnecting n_1, n_2 and n_3 by $e_{n_1 n_2}$ and $e_{n_1 n_3}$. The length difference between the minimum spanning tree and the MRST is determined by the location of s in edge $e_{n_1 n_3}$, formed by edges $e_{n_1 s}$ and $e_{n_3 s}$. When s is located at the mid-point of $e_{n_1 n_3}$, the maximum length difference, equal to the half length of $e_{n_1 n_3}$, is reached. Figures 2.5c and 2.5d show a special case with the maximum length difference between a minimum spanning tree and the MRST for the same node configuration. In this case, the ratio of the length of the MRST and the length of the minimum spanning tree is 2:3. This indicates that the maximum length reduction produced by replacing a minimum spanning tree by an MRST for the same node configuration is 33%. Actually, s in Figure 2.5c can be considered an overlap of two Steiner points: one is connected to n_2 and another to n_4. Therefore, the length difference between the minimum spanning tree in Figure 2.5d and the MRST in Figure 2.5c is twice of that between the minimum spanning tree in Figure 2.5a and the MRST in Figure 2.5b.

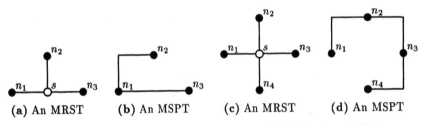

(a) An MRST (b) An MSPT (c) An MRST (d) An MSPT

Figure 2.5. The length difference between an MRST and an MSPT

The average edge length in a minimum spanning tree with the length equal to l_{MSPT} for an n-node configuration is $l_{MSPT}/(n-1)$. According to Corollary 8.3.6 (see Chapter 8), there are at most $n-2$ Steiner points in an MRST for n nodes. If $(n-2)/2$ Steiner points are located on the mid-points of the reconnected edges of the minimum spanning tree, and other $(n-2)/2$ Steiner points superpose with the nodes, $(1/4)l_{MSPT}/(n-1)$ is reduced by increasing a Steiner point. For different node configurations, the average length reduction between a set of minimum spanning trees and a set of MRSTs for n nodes can be estimated by

$$l_{MSPT} - l_{MRST} \approx 0.25 \frac{n-2}{n-1} l_{MSPT} \tag{2.3}$$

For trees with different number of nodes, n is the average number of nodes per tree. Table 2.1 gives the relation between n and $1 - l_{MSPT}/l_{MSPT}$. If the average number of nodes per tree is 6, the total routing length can be reduced 20%.

Table 2.1. The length reduction between an MSPT and an MRST

n	3	4	5	6	8	10	15	20
$1 - \frac{l_{MRST}}{l_{MSPT}}$	0.13	0.17	0.19	0.2	0.21	0.22	0.23	0.24

2.1.4   Chain trees

A spanning tree reduces to a chain if $deg\ n_i = 1$ or 2 $(i = 1, 2, \ldots, |V_n|)$. A node with degree 1 is called an end node (end vertex) of a chain. There are two end nodes in a chain. A set of chains can be expressed as a sequence $\{n_1, n_2, \ldots, n_n\}^*$, in which n_1 is adjacent to n_2, n_2 is adjacent to n_3, ..., and n_{n-1} is adjacent to n_n. The total number of different sets of chains for V_n is $|V_n|!/2$, because the total number of permutations for nodes in V_n is $|V_n|!$, and $\{n_1, n_2, ..., n_{|V_n|}\}^*$ and $\{n_{|V_n|}, ..., n_2, n_1\}^*$ express the same set of chains. The total number of chains expressed by $\{n_1, n_2, ..., n_n\}^*$ is $\prod_{i=1}^{n-1} C(|x_{n_i} - x_{n_{i+1}}| + |y_{n_i} - y_{n_{i+1}}|, |x_{n_i} - x_{n_{i+1}}|)$, where x_{n_i} and y_{n_i} are the coordinates of n_i. A chain of minimum length is called a minimum chain of a particular node configuration. Figure 2.3 is also a minimum chain.

The length of a minimum chain is always equal to or longer than the length of a minimum spanning tree. In addition, the number of possible alternative chains of the same length as the minimum spanning tree is equal to or less than the number of alternative minimum spanning trees for the same node configuration.

2.1.5   Source-sink trees

A spanning tree reduces to a source-sink tree when $deg\ n_s = |V_n| - 1$ and $deg\ n_{s'} = 1$ for each source n_s; $n_{s'} \in V_n - n_s$ is called a sink. A source-sink tree of minimum length is called a minimum source-sink tree of a particular node configuration. Figure 2.6 shows a source-sink tree for the node configuration in Figure 2.2, assuming the source node.

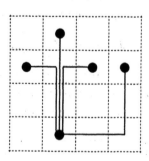

Figure 2.6. A minimum source-sink tree (length = 13 units)

For a given node configuration, the length of a minimum source-sink tree is equal to or longer than that of a minimum spanning tree; the number of alternative source-sink trees of the same length as the minimum spanning tree is equal to or less than the number of alternative minimum spanning trees. Because $n - 1$ nodes of V_n must be adjacent to one node, the constraints on connecting a source-sink tree are much stricter than on a chain. The length of the minimum chain is usually shorter than the length of the minimum source-sink tree, and there are usually more alternative chains than alternative source-sink trees for the same node configuration.

2.1.6 Row-based trees

Approximating a minimum Steiner tree is a time-consuming process. As an alternative, a new tree type, the row-based tree, was developed [WON89], in which all nodes in a row are interconnected, and then the nearest rows are connected. Figure 2.7 shows a row-based tree for the same node configuration as in Figure 2.2. The time needed to compute the length of a row-based tree is linear to the number of its rows. There are at most n rows in a row-based tree that connect n points.

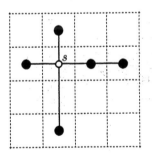

Figure 2.7. A row-based tree

If each end node of every row is restricted to connecting to the nearest row or node, then the row-based tree is called a row-based tree with corrected terminal connections. A row-based tree with corrected terminal connections provides more alternative configurations than the minimum spanning tree (see Chapter 4). Furthermore, a row-based tree with corrected terminal connections more closely approximates an MRST in an iterative placement process. By correcting the connection of each terminal node in every column, the row-based tree can also be developed into a row-column-based tree to approximate even more closely the minimum Steiner tree in placement.

2.1.7   Trees with special edges

In high-speed circuitry, such as that encountered in advanced computer equipment using ECL technology, signals often must propagate with a common propagation time or with a relatively longer (or shorter) propagation time than other signals, or propagate within a certain time limit of other signals to avoid electrical race conditions. In micro-wave circuitry, transmission line effects must also be considered. In some cases, the lengths and shapes of edges (or paths) are specified. In general, an edge or path can be routed according to one of the following requirements:

- The length of an edge or a path is bounded by or within a specified length (Figures 2.8 and 2.9). The specified length may be a set value or another connection. The bound may be a minimum, a maximum or a range around another value.
- The shape of an edge is specified or the shapes of two edges or two paths in a tree are matched (matched pair) (Figure 2.10).

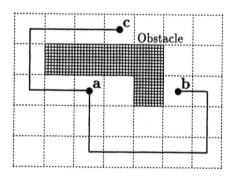

Figure 2.8. Special edges, with $l_{ab} = l_{ac} + 2$, in a minimum spanning tree

Considering obstacle in detailed routing, the minimum Steiner tree is still the best to satisfy special requirements, because it provides the minimum routing length of any tree type, and often the most connection options.

Example: In Figure 2.8, a minimum spanning tree turning around an obstacle is used to determine a path such that edges e_{ab} and e_{ac} are bounded by the relationship $l_{e_{ab}} = l_{e_{ac}} + 2$, where $l_{e_{ab}}$ and $l_{e_{ac}}$ are the lengths of edges e_{ab} and e_{ac} respectively. Figure 2.9 shows a minimum Steiner tree turning around the same obstacle for the same requirements.

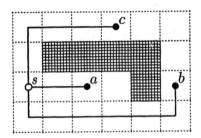

Figure 2.9. Special paths, with $l_{e_{ab}} = l_{e_{ac}} + 2$, in a Steiner tree

◇

Example: Figure 2.10 shows that a matched pair that can form a Steiner tree turning around obstacles. However, a matched pair that form a spanning tree for the same obstacles and the same node configuration cannot be found. This example shows that the Steiner tree provides more options than the spanning tree in finding a matched pair.

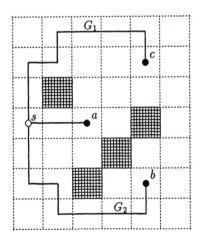

Figure 2.10. A matched pair $(G_1 = G_2)$

◇

2.2 Generating Minimum Spanning and Chain Trees

A minimum tree is the tree in a specific type with the minimum length. Generating minimum trees for a variety of tree types is the foundation of placement for routability and routing. This section discusses the theory for generating minimum spanning trees, and the relationship between graph partitioning and total minimum spanning trees; as well as an approximation method for generating minimum chains.

2.2.1 Generating a minimum spanning tree

A minimum spanning tree must satisfy the following objective function and constraint conditions:

Objective function:

$$l_t = minimizing(\sum_{i=1}^{n-1} \sum_{j=i+1}^{n} d_{v_i v_j} \cdot x_{ij})$$

Constraint conditions:

(1) $x_{ij} = x_{ji} = 0$ or 1

(2) $\sum_{i=1}^{n-1} \sum_{j=i+1}^{n} d_{v_i v_j} \cdot x_{ij} = \sum_{k=1}^{n-1} d_{v_k v_m}$ ($m \neq k$ and $1 \leq m \leq n$)

where the first constraint given by x_{ij} denotes whether or not the edge $v_i v_j$ exists, and the second constraint guarantees that all nodes are connected without cycles. The total number of all different edges is $C(n, 2) = \sum_{i=1}^{n-1} i(n-i) = \frac{1}{2}n(n-1)$, and the total number of all different graphs with $(n-1)$ edges is $C(\frac{1}{2}n(n-1), n-1)$. The minimum spanning trees can be identified by examining all potential graphs under the constraint conditions, but this method is inefficient.

Let $V_n = V_1 \cup V_2$, where $V_1 = \{n_{11}, n_{12}, \ldots, n_{1a}\}$, $V_2 = \{n_{21}, n_{22}, \ldots, n_{2b}\}$, $V_1 \cap V_2 = \emptyset$, and $a + b = |V_n|$. There are $a \times b$ edges in the bipartite graph $G = (V_1, V_2, E)$ (see Figure 2.11), where

$$E = \{ \quad n_{11}n_{21}, \ldots, n_{11}n_{2b},$$

$$\vdots$$

$$n_{1a}n_{21}, \ldots, n_{1a}n_{2b} \}$$

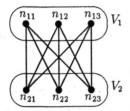

Figure 2.11. Edges crossing two subsets of nodes

Let $n_{1i}n_{2j}$ be the edge connecting $n_{1i} \in V_1$ and $n_{2j} \in V_2$. If $d_{n_{1x}n_{2y}} = min(d_{n_{1i}n_{2j}}; i = 1, 2, \ldots, a; j = 1, 2, \ldots, b)$, then edge $n_{1x}n_{2y}$ is the shortest edge in G.

Theorem 2.1. $\forall e \in E(t)$, if $t - e = \{t_1, t_2\} = \{(V_1, E(t_1)), (V_2, E(t_2))\}$ ($t_1 \cap t_2 = \emptyset$) and e is the shortest edge in the bipartite graph $G = (V_1, V_2, E)$, then t is a minimum spanning tree.

Proof: If $\exists e' \notin E(t)$, $t' = t - e + e'$, and $l_{t'} < l_t$, then $\exists c \subset t + e'$, where c is a cycle and $e, e' \in c$ (see Figure 2.12). Let l_c be the length of the cycle, $l_t - (l_c - l_{e'}) = l_{t'} - (l_c - l_e)$, and therefore $l_{e'} < l_e$. However, this result conflicts with the given condition that e is the shortest edge in $G = (V_1, V_2, E)$, where $\{(V_1, E(t_1)), (V_2, E(t_2))\} = t - e$. It is impossible that $l_{t'} < l_t$. $\square$

Figure 2.12. A shorter path in a cycle

According to Theorem 2.1, trees connecting nodes in V are the minimum spanning trees, if they can be separated by arbitrarily partitioning V into V_1 and V_2, and if each edge of every tree connects two nearest nodes $u \in V_1$ and $v \in V_2$. Corollary 2.1.1 further explains the relationship between partitioning and minimum spanning trees t', t'', ..., and $t^{(n)}$ in placement.

Corollary 2.1.1. If $t^{(i)} = (V^{(i)}, E(t^{(i)}))$ ($i = 1, \ldots, n$), and $u_i v_i \in E(t^{(i)})$ is a shortest edge connecting nodes $u_i \in V_1$ and $v_i \in V_2$, where V_1 and V_2 are arbitrarily partitioned from $V = \cup_{i=1}^{n} V^{(i)}$, then t', t'', ..., $t^{(n)}$ are the minimum spanning trees.

Corollary 2.1.1 indicates that minimizing the total length of the minimum spanning trees is equivalent to minimizing the total length of edges connecting two nodes in different subsets. Corollary 2.1.1 provides an efficient method to minimize the total length of minimum spanning trees in placement, because computing the length of a shortest edge is much more efficient than computing the length of a minimum spanning tree. If the minimum spanning tree is employed in routing, rather than in placement, it must be generated. Using Theorem 2.1, the following procedure finds a minimum spanning tree by partitioning.

> PROCEDURE MinSpanningTree1(V_n);
> BEGIN
> > Arbitrarily partition V_n into V_1 and V_2;
> > Find a shortest edge nn' to connect $n \in V_1$ and $n' \in V_2$;
> > $V_1 = \emptyset$; $E(t) = \emptyset$;
> > WHILE $V_2 \neq \emptyset$ DO BEGIN
> > > $V_1 = V_1 \cup \{n, n'\}$;
> > > $V_2 = V_n - V_1$;
> > > $E(t) = E(t) \cup nn'$;
> > > Find a shortest edge nn' to connect $n \in V_1$ and $n' \in V_2$;
> > END;
> END.

Finding a shortest edge is called a pass in a computational process. A total of $n - 1$ passes are required to find a minimum spanning tree. To determine the shortest edge in a pass, lengths of $i(n - i)$ edges are computed and compared, where $i = |V_1|$. Therefore, the time needed to complete a minimum spanning tree is proportional to $\sum_{i=1}^{n-1} i(n - i) = \frac{1}{6}n(n^2 - 1)$.

If redundant computations in the above procedure can be eliminated, the time needed to find a minimum spanning tree will be reduced. Therefore, instead of implementing the first step in the procedure, a set of different edge lengths

$$\{l_{v_i v_j}; i = 1, 2, \ldots, n - 1; \ j = i + 1, \ldots, n\}$$

is computed, and then all edges are sorted in ascending order of edge lengths.

The total number of all different edges is $\sum_{i=1}^{n-1}(n-i) = n(n-1)/2$, and the total number of comparisons in the sorting is $[n(n-1)-1]log_2\frac{n(n-1)}{2}$. Each order number corresponding to the edge order in the sorted list is placed in an upper or lower triangular matrix whose rows correspond to $v_1, ..., v_{n-1}$ and whose columns correspond to $v_2, ..., v_n$. The order matrix is an adjacent matrix with the order of edge length as elements.

Section 1.3.5 states that a cycle exists if n nodes are connected by n different edges. To determine the cycle, all edges of V_1 in a set of sub-trees are stored, and the address of each node of V_1 associated with the name of the connected sub-tree is written into a table. A cycle exists, if and only if an edge has two vertices common with two nodes in the same sub-tree. The modified procedure for generating a minimum spanning tree uses both the sorted edge-set list and the method of determining cycles.

> *PROCEDURE MinSpanningTree2(V_n);*
>
> *BEGIN*
>
>   *Compute $2(n-1)/2$ edges for the complete graph connecting V_n;*
>
>   *Sort all edges in ascending order;*
>
>   *Store nodes connected by the first and second edges in V_1;*
>
>   *Move the first and second edges to $E(t)$;*
>
> $V_2 = V_n - V_1; i = 3;$
>
>   *WHILE $V_2 \neq \emptyset$ DO BEGIN*
>
>   *IF the ith edge with edges in $E(t)$ form a cycle THEN*
>
> $i = i + 1$ {Skip this edge}
>
>   *ELSE BEGIN*
>
>   *Store nodes connected by the ith edge in V_1;*
>
>   *Move the ith edge to $E(t)$;*
>
> $V_2 = V_n - 1;$
>
> $i = i + 1;$
>
>   *END;*
>
>   *END;*
>
> *END.*

Three additions/subtractions are required to compute an edge length, and one subtraction is required for a comparison. The ratio of the time needed to determine a minimum spanning tree using the modified version to the time needed to determine a minimum spanning tree using the original version is $[\frac{n(n-1)}{2} - 1]log_2\frac{n(n-1)}{2} + \frac{1}{2}n(n-1) \times 3]/[\frac{1}{6}n(n^2-1) \times 4]$. The ratio is 0.366 when $n = 10$. If the edge length is measured by the Euclidean distance, the time needed to find a minimum spanning tree is determined by the number of multiplications and square roots. The ratio is approximately equal to $\frac{1}{2}n(n-1)/\frac{1}{6}n(n^2-1) = \frac{3}{n+1}$.

Example: The connection arrangement of nodes in Figure 2.13a is determined using a minimum spanning tree. Corresponding to the node configuration in Figure 2.13a (see Table 2.2) where $V_n = \{n_i | i = 1, ..., 5\}$, the sorted edge list can be established. An edge with a shorter length is given higher selection priority.

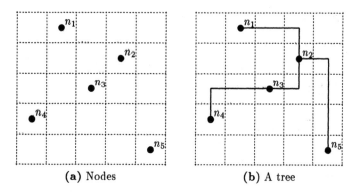

(a) Nodes   (b) A tree

Figure 2.13. A minimum spanning tree

Table 2.2.   A sorted edge list for V_n, as shown in Figure 2.12

Order	1	2	3	4	5	6	7	8	9	10
Edge	n_2n_3	n_1n_2	n_1n_3	n_3n_4	n_1n_4	n_2n_5	n_3n_5	n_2n_4	n_4n_5	n_1n_5
Length	2	3	3	3	4	4	4	5	5	7

According to Table 2.2, the order matrix can be expressed as:

2.2 GENERATING MINIMUM SPANNING AND CHAIN TREES

$$
\begin{array}{c}
\begin{array}{cccc} n_2 & n_3 & n_4 & n_5 \end{array} \\
\begin{array}{c} n_1 \\ n_2 \\ n_3 \\ n_4 \end{array}
\left[
\begin{array}{cccc}
2 & 3 & 5 & 10 \\
  & 1 & 8 & 6 \\
  &   & 4 & 7 \\
  &   &   & 9
\end{array}
\right]
\end{array}
$$

The sorted edge list shown in Table 2.2 demonstrates that $n_2 n_3$ is put on the top selection priority, and therefore $V_1 = \{n_2, n_3\}$, $V_2 = \{n_1, n_4, n_5\}$, and $E(t) = n_2 n_3$. Because the edge of order 1 is removed, this edge in the sorted edge list is not considered in the subsequent processing. Table 2.3 is an updated sorted edge list.

Table 2.3. Removing edge $n_2 n_3$

Order	2	3	4	5	6	7	8	9	10
Edge	$n_1 n_2$	$n_1 n_3$	$n_3 n_4$	$n_1 n_4$	$n_2 n_5$	$n_3 n_5$	$n_2 n_4$	$n_4 n_5$	$n_1 n_5$
Length	3	3	3	4	4	4	5	5	7

In the second pass, $n_1 n_2$ is selected. Therefore, $V_1 = \{n_2, n_3, n_1\}$, $V_2 = \{n_4, n_5\}$ and $E(t) = \{n_2 n_3, n_1 n_2\}$. The sorted edge list is updated when the edge of order 2 is removed (see Table 2.4).

Table 2.4. Removing edge $n_1 n_2$

Order	3	4	5	6	7	8	9	10
Edge	$n_1 n_3$	$n_3 n_4$	$n_1 n_4$	$n_2 n_5$	$n_3 n_5$	$n_2 n_4$	$n_4 n_5$	$n_1 n_5$
Length	3	3	4	4	4	5	5	7

Because the edge of order 3 forms a cycle with the edges of order 1 and 2, $n_1 n_3$ is ignored. In the third pass, $n_3 n_4$ is selected, because it cannot form a cycle with the previously selected edges. Therefore, $V_1 = \{n_2, n_3, n_1, n_4\}$, $V_2 = n_5$ and $E(t) = \{n_2 n_3, n_1 n_2, n_3 n_4\}$. The sorted edge list is again updated (see Table 2.5).

Table 2.5. Ignoring $n_1 n_3$ and removing edge $n_3 n_4$

Order	5	6	7	8	9	10
Edge	$n_1 n_4$	$n_2 n_5$	$n_3 n_5$	$n_2 n_4$	$n_4 n_5$	$n_1 n_5$
Length	4	4	4	5	5	7

Because the edge of order 5 forms a cycle with the edges of order 3 and 4, $n_1 n_4$ is removed. In the fourth pass, $n_2 n_5$ is selected, $V_1 = \{n_2, n_3, n_1, n_4, n_5\}$, $V_2 = \emptyset$ and $E(t) = \{n_2 n_3, n_1 n_2, n_3 n_4, n_2 n_5\}$. The tree t is shown in Figure 2.13b.

◇

2.2.2   Generating a minimum chain tree

Finding a minimum chain may intuitively involve solving the traveling salesman problem, but without return to the original city. A minimum chain doesn't necessarily correspond to the shortest cycle, nor is the cycle which is generated by connecting two end-nodes of a minimum chain necessarily the shortest cycle.

Example: In Figure 2.14d, the cycle is shortest, but the corresponding chain is not. In Figure 2.14c, the cycle obtained by connecting two end-nodes of the minimum chain is not the shortest. However, Figures 2.14a and 2.14b show both the chains and their corresponding cycles are the shortest.

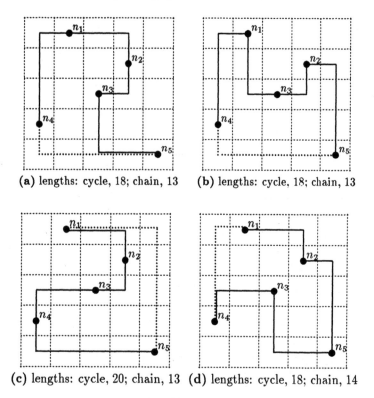

(a) lengths: cycle, 18; chain, 13   (b) lengths: cycle, 18; chain, 13

(c) lengths: cycle, 20; chain, 13 (d) lengths: cycle, 18; chain, 14

Figure 2.14. A minimum chain is related to an end-node

To find minimum chains by the recursive search used in the traveling salesman problem, some modifications are required:

- The length between two end nodes is omitted from the cycle length.
- Each node is assumed to be an end node, so that a local minimum chain corresponding to this end node can be found.

The global minimum chains are determined from the local minimum chains corresponding to the assuming end nodes. For every given end node, there is a corresponding local minimum chain. A global minima can be found from n local minima. Therefore, the time needed to determine a minimum chain is n times that of determining a minimum cycle, using the recursive search of the traveling salesman problem.

Example: If node n_5 is considered to be an end-node, a minimum chain can be obtained from a recursive search similar to that employed in solving the traveling salesman problem (Figures 2.14a, 2.14b and 2.14c).

◇

Because minimum chains is searched from the complete graph that connects all given nodes, a simplified search process for a minimum chain follows:

- Find a simplified graph, G_e, which is obtained by removing the maximum number of edges from the complete graph, but contains at least one minimum chain.
- Determine an end node for a minimum chain.

In complete graph G_c for $V_n = \{n_1, n_2, \ldots, n_n\}$, $deg\ n_i = n - 1$ $(1 \leq i \leq n)$. When edges are removed from G_c, the degree of an end node of the removal edge is reduced. If E_i is a set of edges with the same length in G_c, and the length of an edge in E_i is l_{E_i}, a sorted edge list, E_c for G_c, can be expressed as the following sequence:

$$E_c = \{E_1, E_2, \ldots, E_m\}^* \tag{2.4}$$

where $l_{E_1} < l_{E_2} < \ldots < l_{E_m}$ and $m < n$.

To minimize the length of a chain, edges in the chain can be selected from the complete graph E_c, until all edges for the chain are found. Because the degree of a node in a chain is one or two, G_e can be approximated by initially removing the longest edges, E_m, from G_c, then removing shorter edges, until an edge with an end node of degree two is the next to be removed. Note that the degree of each node in G_e is greater than 1.

Theorem 2.2. If $deg\, n_i = 2$ and $n_i n_j \in E(G_e)$ has the longest length, then n_i is an end-node of a chain with the minimum length of all chains contained in G_e.
Proof: Let $c \subset G_e$ be a cycle connecting all nodes and $\{n_i n_j, e\} \subset c$. If $t = c - n_i n_j$ and $t' = c - e$, then $l_t \leq l_{t'}$, because $n_i n_j$ is the longest edge in $E(G_e)$. A chain which contains $n_i n_j$ must not be the minimum chain of all chains contained in G_e. If $n_i n_j$ is removed from G_e, then $deg\, n_i = 2 - 1 = 1$. Therefore, not only must n_i be an end-node of a chain without $n_i n_j$ as an edge, but it is also an end node of a chain of the minimum length contained in G_e. □

The time needed to find minimum chains from G_e is dominated by the number of edges in G_e. When complete graph G_c is reduced to graph G_e, the minimum degree is changed from $n - 1$ to 2. The minimum difference between the edge number of G_c and the edge number of G_e is $n - 3$, and the maximum difference is $n(n-1)/2 - (n-1) = (n-1)(n-2)/2$. Two nodes exist in the second level of the search tree for the minimum chain, rather than the $n - 1$ nodes in the second level of the search tree for the traveling salesman problem. In addition, when G_e is obtained by removing $n - 3$ edges from G_c, the time needed to complete the recursive search for a minimum chain is less than $2/(n - 1)$ the time needed to complete the recursive search for a cycle.

When G_e is obtained by removing $1 + (n-1)(n-2)/2$ edges from G_c, G_e is a cycle; the time needed to complete the recursive search for the traveling salesman problem can be saved in searching for the minimum chain, because a minimum chain can be immediately obtained by removing the longest edge in the cycle.

Example: The node configuration in Figure 2.15 is used to explain how to reduce graph G_c to graph G_e in which minimum chains can be found. Like the sorted edge list used in finding a minimum spanning tree, $E_c = \{E_1, ..., E_5\}^*$ can be expressed as the following sorted edge list:

Table 2.6. A sorted list for finding a minimum chain

	E_1	E_2			E_3			E_4		E_5
Edge	$n_2 n_3$	$n_1 n_2$	$n_1 n_3$	$n_3 n_4$	$n_1 n_4$	$n_2 n_5$	$n_3 n_5$	$n_2 n_4$	$n_4 n_5$	$n_1 n_5$
Length	2	3	3	3	4	4	4	5	5	7

The complete graph, G_c, which corresponds to the sorted edge list given in Table 2.6, is shown in Figure 2.15a, where the node configuration is the same as in

Figure 2.14. After removing $E_5 = n_1n_5$ and $E_4 = \{n_2n_4, n_4n_5\}$ from G_c, further removing an edge in $E_3 = \{n_1n_4, n_2n_5, n_3n_5\}$ causes $deg\ n_4 = 1$ or $deg\ n_5 = 1$. Therefore, the simplified graph, G_e, is found (see Figure 2.15b). According to Theorem 2.2, n_4 or n_5 can be selected as an end-node.

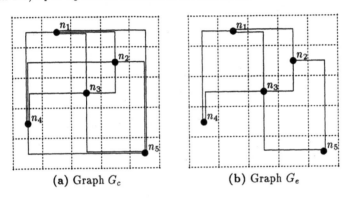

(a) Graph G_c (b) Graph G_e

Figure 2.15. Finding end-nodes for minimum chains

$\diamond$

Example: Figure 2.16 is the search tree for the minimum chain, which is obtained by selecting n_4 as an end node and removing n_1n_4 from G_e in Figure 2.15b. In this search tree, n_5^* and n_1^* express $n_1n_5 \notin E(G_e)$. Chain $\{n_4, n_3, n_1, n_2, n_5\}^*$ is a minimum chain with a length of 13 units. Chain $\{n_4, n_3, n_1, n_2, n_5\}^*$ shows that the shortest edge may not be contained in a minimum chain. However, most of the edges in the minimum chain belong to those in E_1 and E_2. Similarly, a minimum chain $\{n_4, n_1, n_3, n_2, n_5\}^*$ (see Figure 2.14b) can be found by selecting n_1 as the second node in a search tree, with n_4 as the end node.

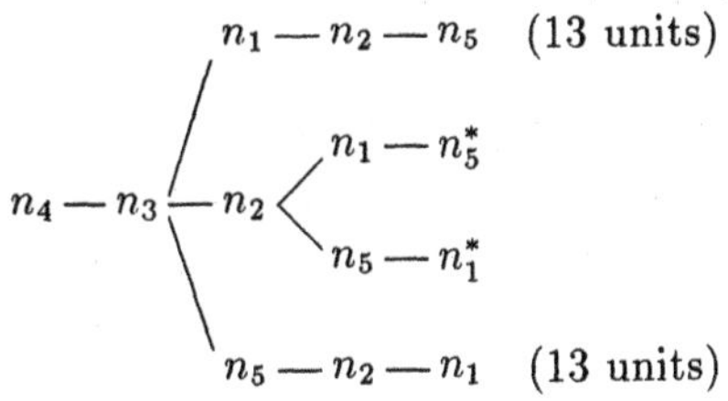

$$n_1 - n_2 - n_5 \quad (13 \text{ units})$$

$$n_4 - n_3 - n_2 \left\langle \begin{array}{l} n_1 - n_5^* \\ \\ n_5 - n_1^* \end{array} \right.$$

$$n_5 - n_2 - n_1 \quad (13 \text{ units})$$

Figure 2.16. The search tree with n_4 as an end node

$\diamond$

Example: When n_5 and n_2 are selected as an end node and the second node, respectively, removing n_5n_3 from Ge in Figure 2.15b gives the search tree shown as Figure 2.17. This search tree gives three minimum chains with a length of 13 units: $\{n_5, n_2, n_1, n_3, n_4\}^*$, $\{n_5, n_2, n_3, n_1, n_4\}^*$ and $\{n_5, n_2, n_3, n_4, n_1\}^*$.

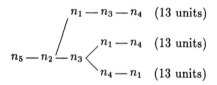

Figure 2.17. The search tree with n_5 as the end-node
and n_2 as the second node

The search tree shown in Figure 2.18 is obtained by removing $n_5 n_2$ and selecting n_5 as an end-node and n_3 as the second node. Only one minimum chain, $\{n_5, n_3, n_2, n_1, n_4\}^*$, can be found from this search tree.

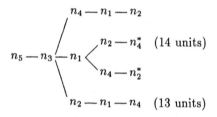

Figure 2.18. The search tree with n_5 as the end-node
and n_3 as the second node

◇

2.3   Minimum Steiner Tree Approximations

Find a minimum Steiner tree from the combinations of all possible Steiner trees requires the problem be divided into sub-problems in which the local minima are found from topologies, possible connection relations among the given nodes and a set of given Steiner points. Gilbert-Pollak's formula [GIL68] demonstrates that the number of possible topologies is $N_t(n) = \sum_{n_s=0}^{n-2} 2^{-n_s} C(n, n_s+2)[(n+n_s-2)!]/(n_s!)$, where n_s $(0 \leq n_s \leq n - 2)$ is the number of Steiner points, and n is the number of nodes. Table 2.7 gives the number of possible topologies when $3 \leq n \leq 10$ and shows that it is extremely difficult to find an MRST from topologies for more than 5 nodes.

Table 2.7. The number of possible topologies ($3 \leq n \leq 10$)

n	3	4	5	6	7	8	9	10
$N_t(n)$	4	27	270	3,645	62,370	1,295,595	30,717,614	652,017,144

Finding an MRST is an NP-complete problem [GAR77]. Before a theory solving the Steiner's problem is developed, traditionally, an MRST is approximated by connecting MRSTs for three nodes, or replacing longer edges with shorter edges in a minimum spanning tree, or restricting the number of nodes (typically less than 6), or locating nodes on a particular configuration (typically the perimeter of a rectangle) [CHA72, HAN72, SOU75, LEE76, AHO77, HWA79, ANT86, CHA90, COH90, KAH90, HO90, KAH92].

2.3.1 An approximation based on MRSTs for three nodes

For simplicity, approximations of MRSTs for a tree with $|V_n| = 3$ are considered first. Many methods can be employed to find an MRST for three nodes. According to Theorem 8.3, the Steiner point of an MRST for three nodes can be found by a very efficient computation. For $V_n = \{n_1, n_2, n_3\}$, in which $n_1 = (x_{n_1}, y_{n_1})$, $n_2 = (x_{n_2}, y_{n_2})$ and $n_3 = (x_{n_3}, y_{n_3})$, Steiner point $s = (s_x, s_y) = (mid(x_{n_1}, x_{n_2}, x_{n_3}), mid(y_{n_1}, y_{n_2}, y_{n_3}))$, where mid is the function of selecting the mid-value from three given values. Therefore, the procedure of finding an MRST for $V_n = \{n_1, n_2, n_3\}$ can be given as follows:

PROCEDURE $t(n_1, n_2, n_3)$;

BEGIN

   $x_s = mid(x_{n_1}, x_{n_2}, x_{n_3})$;

   $y_s = mid(y_{n_1}, y_{n_2}, y_{n_3})$;

   IF $s = (x_s, y_s) \notin V_n$ *THEN BEGIN*

      $V_s = s$;

      $E(t) = \{e_{n_1 s}, e_{n_2 s}, e_{n_3 s}\}$;

   END

   ELSE IF $s \notin \{n_i, n_j\}$ $(1 \leq i \neq j \leq 3)$ *THEN* $E(t) = \{e_{n_i s}, e_{n_j s}\}$;

   $t = (V_s \cup V_n, E(t))$;

END.

Example: This procedure's implementation is shown in Figure 2.19: (a) is the case with one Steiner point, and (b) with no Steiner point, or a Steiner point that overlaps with a node.

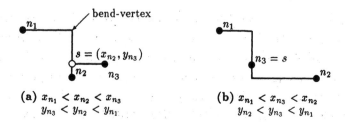

$$(\text{a}) \ x_{n_1} < x_{n_2} < x_{n_3} \qquad\qquad (\text{b}) \ x_{n_1} < x_{n_3} < x_{n_2}$$
$$y_{n_3} < y_{n_2} < y_{n_1} \qquad\qquad\qquad y_{n_2} < y_{n_3} < y_{n_1}$$

Figure 2.19. Stainer point $s = (mid(x_{n_1}, x_{n_2}, x_{n_3}), mid(y_{n_1}, y_{n_2}, y_{n_3}))$

◇

A tree approximating an MRST with a set of arbitrary nodes can be obtained by connecting an unconnected node with two connected nodes into an MRST for three nodes. If $x_{n_1} \neq x_{n_2}$ and $y_{n_1} \neq y_{n_2}$, then there is a bend in edge $n_1 n_2$. The computation of finding a Steiner point can be simplified when an edge is split on the bend-vertex and stored as straight edges (see Figure 2.19). As proposed by J. H. Lee, N. K. Bose and F. K. Hwang [LEE76], the procedure can be expressed:

> PROCEDURE RST(V_n);
>
> BEGIN
>
>   Find $C(|V_n|, 3)$ different MRSTs with three nodes;
>
>   Denote one with the minimum length as $t = (V(t), E(t))$, in which
>
> $V(t) = V'_n \cup V'_s$ and $|V'_n| = 3$;
>
>   IF there is a bend vertex b in an edge $e_{v_1 v_2} \in E(t)$ THEN BEGIN
>
> $E(t) = (E(t) - e_{v_1 v_2}) \cup \{e_{v_1 b}, e_{v_2 b}\}$;
>
>   {Separate a bend into a horizontal and a vertical edges}
>
> $V'_s = V'_s \cup b$; {Collect b as a "Steiner point"}
>
>   END;
>
>   WHILE $|V'_n| \neq |V_n|$ DO BEGIN
>
>   Find nodes $v \in V_n - V'_n$ and edge $e_{v'v''} \in E(t)$ to minimize $l_{e_{vs}}$;
>
>   {s is determined by v, v', and v''};
>
>   IF $s \notin V(t)$ THEN BEGIN

$$V'_s = V'_s \cup s;$$
$$E(t) = E(t) \cup e_{vs};$$

END

ELSE IF a bend vertex $b \in e_{vs}$ *THEN BEGIN*

$$E(t) = E(t) \cup \{e_{sb}, e_{bv}\};$$
$$V'_s = V'_s \cup b;$$

END

ELSE $E(t) = E(t) \cup e_{vs};$

$$V'_n = V'_n \cup v;$$

END; {WHILE}

$$RST = (V_n \cup V_s, E(t));$$

END.

This algorithm needs $O(|V_n|^3)$ operations because $|V_n|(|V_n| - 1)(|V_n| - 2)/6$ MRSTs for three nodes are found in the initial state. If the number of MRSTs for three nodes can be reduced in determining the initial MRST for three nodes, the number of operations can be reduced.

2.3.2 Staircase layouts

Arranging modules and interconnecting electronic nets (trees) on a substrate is called layout. The layout of a shortest rectilinear edge, $e_{p_1 p_2}$, between nodes p_1 and p_2, is called a staircase layout of the edge. Merging redundant segments that overlap with other edges in a minimum spanning tree (MSPT), results in a rectilinear Steiner tree (RST). Approximating the MRST, J. Ho, G. Vijayan and C. K. Wong [HO90] presented staircase layouts in order to obtain an MSPT with the maximum redundant segment length. A staircase layout of an edge having at most one or two bends (turns) is called an *L-shape layout* or a *Z-shape layout*, respectively.

Example: Figure 2.20a demonstrates the interconnections in a minimum spanning tree. Figure 2.20b shows all staircase layouts of the minimum spanning tree, where $e_{p_1 p_6}$ and $e_{p_3 p_4}$ are L-shape layouts, and $e_{p_1 p_2}$ and $e_{p_3 p_5}$ are Z-shape layouts.

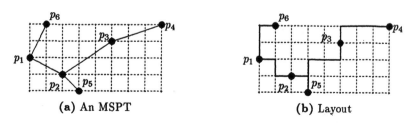

(a) An MSPT   (b) Layout

Figure 2.20. A minimum spanning tree and its edge layout

◇

An RST obtained from an MSPT with the staircase layout is denoted as S-RST, an RST obtained from an MSPT with the L-shape layout is denoted as L-RST, and an RST obtained from an MSPT with the Z-shape layout is denoted as Z-RST. Replacing a staircase segment with another staircase segment between the same end-vertices is called *staircase re-routing*. Staircase re-routing may cause redundant segments. Merging the redundant segments results in a new S-RST with shorter length.

Example: Figure 2.21 shows an MSPT of a set of eight nodes $\{p_1, p_2, ..., p_7, p_8\}$, and two different S-RSTs of the MSPT. The points s_1, s_2, s_3 and s_4 shown in Figures 2.20a and 2.20b are Steiner points. The length of S-RST Ψ_1 is shorter than that of S-RST Ψ_2. Re-routing any staircase in Ψ_1 does not cause redundant segments. However, the staircase segment between s_1 and p_3 in Ψ_2 can be re-routed to cause redundant segments in the staircase segment between p_3 and p_4. When this redundant segment is merged, the resulting S-RST has the same length as Ψ_1.

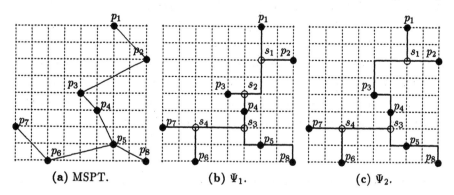

(a) MSPT.   (b) Ψ_1.   (c) Ψ_2.

Figure 2.21. An MSPT and two S-RSTs

◇

An S-RST is said to be stable under re-routing if the staircase re-routing of any subset of its staircase segment dose not cause any redundant segments. In Figure 2.21, Ψ_1 is stable and Ψ_2 is not. An optimal S-RST must be stable. Ho *et al* proved that an optimal Z-RST is also an optimal S-RST. However, optimal L-RSTs do not exhibit stability. An MSPT employed in the L-shape layout must have a characteristic called separability: in any pair of non-adjacent edges in an MSPT, any staircase layouts of those two edges should not intersect or overlap. A separable MSPT is denoted as SMSPT. Note that an SMSPT may not be an MSPT.

Example: In Figure 2.22, (a) demonstrates a non-separable MSPT, and (b) demonstrates a SMSPT, but not an MSPT.

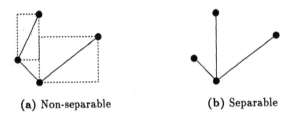

(a) Non-separable   (b) Separable

Figure 2.22. A non-separable MSPT and a separable MSPT

◇

The time complexity of finding an MSPT is approximately $O(n^2)$ for a set of n given nodes. If an edge in an SMSPT is deleted, two generated sub-trees cannot intersect or overlap each other. A redundant segment can only occur in the overlap of two adjacent edges. Algorithms for optimal L-RST and for optimal Z-RST can be obtained by using SMSPT.

Let $\{e_1, e_2, ..., e_n\}$ be the set of edges in a SMSPT. Because at most two possible L-shape layouts exist for any edge in the SMSPT, and because redundant segments may exist only in two adjacent edges, the time needed to find an optimal L-RST is linear with the number of edges in the SMSPT. The procedure can be expressed as below.

PROCEDURE $t(e_1, ..., e_n)$;

BEGIN

   For i=1 TO n DO

IF *overlaps exist between* e_i *and its adjacent edges* THEN BEGIN

Fix *the L-shape layout of* e_i *generating longer overlap;*

Merge *the redundant segment in* t;

END;

END.

Finding an optimal Z-RST is more complex. However, the experimental result shows that the tree length difference between an optimal L-RST and an optimal Z-RST is not significant.

2.3.3   Nodes lying on a rectangle perimeter

Placing a given set of nodes, V_n, on the perimeter of a rectangle typically fits the terminal configuration of a switch-box. Here, introduced is a linear-time algorithm proposed by J. P. Cohoon, D. S. Richards and J. S. Salowe [COH90].

A set of nodes lies on rectangle boundary B. If T is the set of MRSTs connecting the given nodes and each tree in T is composed of a set of horizontal and vertical line segments, s, then T must be enclosed by B. For tie-breaking (interconnection breaking) rules, the lower left corner of the rectangle is located at the origin.

The algorithm introduces three *tie-breaking rules: exterior wire, leftness and topness.* The exterior tie-breaking rule prefers Steiner trees in $T_1 = \{s|s \in t' \in T_1, l_{s\cap B}$ *is maximized*\}, where t' is a Steiner tree connecting V_n and l_x is the total length of segments in x. This rule maximizes the use of the tie segments in the boundary.

The leftness tie-breaking rule prefers Steiner trees in $T_2 = \{s|s \in T_1, \sum \tau l_{s\cap(x=\tau)}$ *is minimized*\}, where $x = \tau$ is a vertical line and $\sum \tau l_{s\cap(x=\tau)}$ is called the leftness. Using this rule, ties in T_1 are broken by choosing wire segments of a Steiner tree in T_2 as far to the "left" as possible.

Similarly, the topness tie-breaking rule prefers Steiner trees in $T_3 = \{s|s \in T_2, \sum \tau l_{s\cap(y=\tau)}$ *is maximized*\}, where $y = \tau$ is a horizontal line, and $\sum \tau l_{s\cap(y=\tau)}$ is called the topness. Using this rule, ties in T_2 are broken by choosing wire segments of a Steiner tree in t_2 as far to the "top" as possible. Then $t \in T_3$ is an MRST.

The topology of Steiner trees, consists of interior lines and interior vertices. An interior line is a maximum sequence of adjacent, collinear edges inside the rectangular boundary. A corner vertex is a point in the interior of the rectangle incident to exactly one horizontal edge (leg) and exactly one vertical edge (leg). If both legs of a corner intersect B, the corner is called a complete interior corner. A T-vertex is an interior point incident to precisely three edges, in which the two collinear edges are called the head of T, and another edge is called the body of T. A complete interior line is an interior line with end-vertices lying on opposite sides of B.

J. P. Cohoon, *et al*, proved that the interior lines of an MRST must have one of the following ten topologies, which can be divided into five types:

- boundary tree with no interior edges (Figure 2.23a).
- interior lines with exactly one complete interior line (Figure 2.23b).
- cross with two intersecting interior lines (Figure 2.23c).
- earthworms with two or more parallel complete interior lines (Figure 2.23d).
- complete interior corners in that an odd number of interior lines incident to one leg and exactly one interior line incident to the other leg (Figure 2.23e).

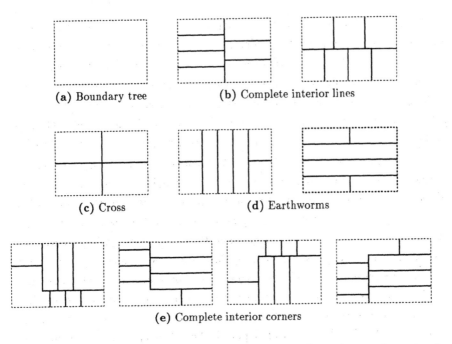

(a) Boundary tree   (b) Complete interior lines

(c) Cross   (d) Earthworms

(e) Complete interior corners

Figure 2.23. Topologies for an MRST with all nodes on the perimeter of a rectangle

Based on the five topologies, a linear-time algorithm for each case can be presented by testing the five types of topologies with exterior wire, leftness and topness. The exterior tie-breaking rule prefers Steiner trees in T_1.

Let $B(a, b)$ be the portion of the boundary (perimeter) traversed from nodes a to b clockwise, and $GAP(a, b)$ be $B(a, b)$ with the maximum interval between two adjacent points removed, $GAP(a,b)$ can be determined by $FUNCTION\ GAP(a, b)$:

> *FUNCTION GAP(a,b);*
>
> *BEGIN*
>
>   *Sort nodes on B in clockwise order;*
>
>   *Save them in a double-linked list;*
>
>   *From a to b find an edge e_{cd} with the maximum length;*
>
>   *GAP = B(a, b) $-e_{cd}$*
>
> *END;*

In this function, c and d are two adjacent nodes in the double-linked list, and $B(a, b)$ is the total length of edges from a to b. The boundary tree can be determined by $FUNCTION\ GAP(a, a)$ in which $a \in V_n$ (see Figure 2.23a).

Based on the boundary tree, the topology with a complete interior line and with no incident interior line can be tested by sliding the complete sinterior line so that it extends from a to b, where at least one of a and b is a terminal. The rest of the tree reduced from the boundary tree must be $GAP(a, b)$ and $GAB(b, a)$.

The difference between the length of the edge and the length of the adjacent interior line can be determined by moving every complete interior line and every incident interior line with at least one end-vertex on B from the right to left or from the bottom to the top along an edge on the boundary. If the length of an edge on the boundary is longer than that of the interior line adjacent to the edge, then the edge is removed. An MRST can be obtained by testing and replacing longer edges with shorter edges in each topology shown in Figures 2.23a to 2.23e. Because the number of comparisons for making decisions to remove edges is proportional to the number of nodes, the time complexity of the algorithm for finding an MRST with all nodes lying on the boundary of a rectangle is linear to the number of given nodes.

2.4 References

[AHO77] Aho, A. V., Garey, M. R., and Hwang, F. K., "Rectilinear Steiner Trees: Efficient Special Case Algorithms," *Networks*, vol. 7, 1977, pp. 37-58.

[ANT86] Antony, P-C. Ng, Raghavan, P., and Thompson, C. D., "A Language for Describing Rectilinear Steiner Tree Configurations," *Proc. 23rd Design Automation Conf.*, 1986, pp. 659-662.

[CHA72] Chang, S., "The Generation of Minimal Trees with a Steiner Topology," *J. Assoc. Comput. Mach.*, vol. 1, 1972, pp. 699-711.

[CHA90] Chao, T-H., and Hsu, Y-C., "Rectilinear Steiner Tree Construction by Local and Global Refinement," *Proc. IEEE Int. Conf. on CAD*, 1990, pp. 432-435.

[COC72] Cockayne, E. J., and Schiller, D. G., "Computation of Steiner Minimal Trees," in *Combinatorics*, D. J. A. Welsh, and D. R. Woodall, eds, Institute for Mathematics and Applications, Southend-on-Sea, Essex, England, 1972, pp. 53-71.

[COH90] Cohoon, J. P., Richards, D. S., and Schowe, J. S., "An Optimal Steiner Tree Algorithm for a Net Whose Terminals Lie on the Perimeter of a Rectangle," *IEEE Trans. on Computer Aided Design*, vol. 9, no. 4, 1990, pp. 398-407.

[COU41] Courant, R., and Robbins, H., *What is Mathematics?*, Oxford University Press, New York, 1941.

[DRE72] Dreyfus, S. E., and Wagner, R. A., "The Steiner Problem in Graphs," *Networks*, 1972, pp. 195-207.

[GAR77] Garey, M. R., and Johnson, D. S., "The Rectilinear Steiner Tree Problem is NP-Complete," *SIAM Journal of Applied Math.*, vol. 32, 1977, pp. 826-834.

[GIL68] Gilbert, E. N., and Pollak, H. O., "Steiner Minimal Trees," *SIAM Journal of Applied Math.*, vol. 16, no. 1, 1968, pp. 1-29.

[HAK72] Hakimi, S. L., "Steiner's Problem in Graphs and Its Implications," *Networks*, vol. 1, 1972, pp. 113-135.

[HAN66] Hanan, M., "On Steiner's Problem with Rectilinear Distance," *SIAM Journal of Applied Math.*, vol. 14, no. 3, 1966, pp. 255-265.

[HAN72] Hanan, M., "A Counter Example to a Theorem of Fu on Steiner's Problem," *IEEE Trans. on Circuit Theory*, CT-19, 1972, pp. 74.

[HO90]   Ho, J. M., Vijayan, G., and Wong, C. K., "New Algorithm for the Rectilinear Steiner Tree Problem," *IEEE Trans. on Computer Aided Design*, vol. 9, no. 2, 1990, pp. 185-193.

[HWA76]   Hwang, F. K., "On Steiner Minimal Trees with Rectilinear Distance," *SIAM Journal of Applied Math.* vol. 30, no. 1, 1976, pp. 104-114.

[HWA79]   Hwang, F. K., "An $O(nlog)$ Algorithm for Suboptimal Rectilinear Steiner Trees," *IEEE Trans. on Circuit and System*, vol. CAS-26, no. 1, 1979, pp. 75-77.

[KAH90]   Kahng, A. B., and Robins, G., "A New Class of Steiner Tree Heuristic with Good Performance: the Iterated 1-Steiner Approach," *Proc. IEEE Int. Conf. on CAD*, 1990, pp. 428-431.

[KAH92]   Kahng, A. B., and Robins, G., "A New Class of Iterative Steiner Tree Heuristics with Good Performance," *IEEE Trans. on Computer Aided Design*, vol. 11, no. 7, 1992, pp. 893-902.

[KRU56]   Krushal, J. B., "On the Shortest Spanning Subtree of a Graph," *Proc. Amer. Math. Soc.*, vol. 7, 1956, pp. 48-50.

[KUH86]   Kuh, E. S., and Marek-Sadowska, M., "Global Routing," in T. Ohtsuki, ed., *Layout Design and Verification*, North-Holland, Amsterdam, 1986, pp. 169-198.

[LEE76]   Lee, H. J., Bose, N. K., and Hwang, F. K., "Use of Steiner's Problem in Suboptimal Routing in Rectilinear Metric," *IEEE Trans. on Circuit and System*, vol. CAS-23, no. 7, 1976, pp. 470-476.

[MEL61]   Melzak, Z. A., "On Problem of Steiner," *Canadian Math. Bulletin*, vol. 4, 1961, pp. 143-148.

[SOU75]   Soukup, J., "On Minimum Cost Networks with Non-linear Costs," *SIAM J. of Appl. Math.*, vol. 25, 1975, pp. 571-581.

[WON89]   Wong, Y. T., and Pecht, M., "Approximating the Steiner Tree in the Placement Process," *ASME J. of Electronics Packaging*, Sept. 1989, pp. 228-235.

Chapter 3

Signal Layer Estimation

Sudha Balakrishnan and Michael Pecht

A typical electronic package consists of devices mounted on a substrate. The substrate, which provides mechanical support and acts as an electrical interconnect medium for the devices, consists of single or multiple layers of etched metallization separated by dielectric that acts as an insulator. Connections between the various layers are made by plated through holes and vias.

The three types of layers commonly observed in multilayer substrates are: signal, power and ground. The power and ground layers are either solid metallization planes or made of traces with sufficient cross-sectional area to support current and voltage requirements. Signal layers support signal interconnections, often with a given layer having traces in either the horizontal or vertical direction for ease of design and manufacture. In this chapter, the focus is on the estimation of the number of signal layers as a function of the magnitude and distribution of the signal circuitry on the substrate.

Typically, an initial estimate of the number of signal layers is made after device placement and prior to routing. The actual number of signal layers is obtained after the routing process, when the interconnection pattern of the traces is defined. However, determining the number of signal layers at such a late stage in the design process can result in undesirable or unmanufacturable products.

Layer estimation at the commencement of the design process provides designers with preliminary cost estimates of the product. For example, the layer-price

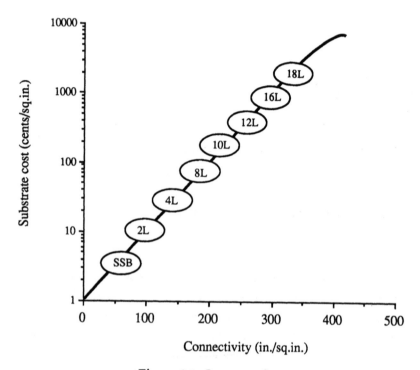

Figure 3.1. Layers vs. Cost

graph in Figure 3.1, can be used to predict substrate cost as a function of the number of layers. Another advantage of a pro-active layer estimation is to appraise conductor dimensions such as, trace width and pad diameter, during substrate design. Finally, most automated routing packages require an estimate of the number of layers prior to the commencement of routing. The more accurate the initial estimation, the faster the routing and shorter the design lead times.

The models presented in this chapter can be used after device selection and prior to device placement during the substrate design process. The inputs available for the estimation process are device configuration, substrate dimensions, conductor dimensions, and interconnection informations (see Table 3.1). The output is the estimated number of signal layers. The models also provide a tool to analyze the effect of design parameters, such as conductor dimensions, routable substrate area, percentage of signal interconnections and device configuration, on the number of signal layers.

Table 3.1. Input and output for the layer estimation

Input:	Output:
• Device configuration:	• Number of signal layers
- Package style	• Effects of
- Device length and width	- conductor dimensions
- Number of leads and lead pitch	- percentage of signal interconnections
• Substrate dimensions:	- routable substrate area
- Substrate length and width	- device configuration on each layer
- Routable substrate area	on the number of single layers
• Conductor dimensions:	
- Via-pad diameter	
- Pitch between two vias	
• Interconnection information:	
- Average net size	
- Number of signal interconnections	

The two main approaches currently in use for estimating the number of signal layers are the density approach and the connectivity approach. Both approaches concentrate on estimating the number of signal layers by evaluating the wiring demand of devices. A uniform congestion, where wires are evenly spread on the surface, is assumed because the topography or approximate location of the interconnections is not known prior to device placement.

Section 3.2 discusses the various factors that affect the number of signal layers. Section 3.3 presents the density approach, where the equivalent integrated circuit count method, used for layer estimation in printed circuit boards, is described. Section 3.4 presents the connectivity approach which can be used for layer estimation in all substrates.

3.1 Factors Affecting Layer Estimation

The primary factors that influence the number of signal layers are routable substrate area, percentage of signal interconnections, device configuration, conductor dimensions, and routing efficiency. This section overviews each of these factors.

Prior to layer estimation, the total device area and the routable substrate area must be compared to determine whether the selected devices can be accommodated on the substrate. Routable substrate area is the surface area that can be used for conductors, including traces, vias, via-pads and device lead pads. Areas for cutouts, clamping, heat sinks and tooling holes are generally excluded from the total substrate area to obtain the routable substrate area. Thus,

$$A_r = (L_b - T_{se})(W_b - T_{se}) - A_{ct} - A_{th} \qquad (3.1)$$

where A_r is the routable substrate area, L_b and W_b are the length and width of the substrate, T_{se} is the trace to substrate edge spacing, A_{ct} is the cutout area, and A_{th} is the tooling hole area. The total device area, A_d, is given by:

$$A_d = \sum_{i=1}^{N_d}(L_{di}W_{di}) \qquad (3.2)$$

where L_{di} and W_{di} denote the length and width of the ith device and N_d is the number of devices. The following conditions, must be satisfied to accommodate devices on the substrate,

$$\begin{aligned} A_d &< A_r \quad \text{for single-sided placement} \\ A_d &< 2A_r \quad \text{for double-sided placement} \end{aligned} \qquad (3.3)$$

If the above conditions are not satisfied, then the substrate or device dimensions must be modified. In general, the number of signal layers is inversely proportional to the routable substrate area. As the routable surface area increases, more area is available per device for routing, and hence the required number of signal layers decreases.

The percentage of signal interconnections is another factor that influences the number of signal layers. On a typical substrate, not all the device connections may be used for signal interconnections. Depending on the magnitude of the signal circuitry, the number of signal interconnections ranges from 30 to 80 percent of the number of device connections. The number of signal layers is directly proportional to the percentage of signal interconnections.

Conductor dimensions include trace width, trace spacing, via-pad diameter and pitch between two vias. When the conductor dimensions are reduced, more traces can be accommodated in a given area, thereby increasing the wiring capacity. The conductor dimensions must be kept as low as possible to achieve the minimum number of layers, but manufacturing constraints restrict the lower limit of conductor dimensions.

Device configuration includes package style (through-hole or surface-mount for components; wire-bonded, tape-automated-bonded and flip-chip for multichip modules and hybrids), number of leads, and lead pitch, each of which affect the number of signal layers. For example, in surface-mount devices the lead pads are

are restricted only to the mounting layer and the corresponding pad area in other layers can be used for routing. In contrast, the pad area of a through-hole component extends through all layers, thereby reducing the effective area available for routing.

Routing efficiency is the ability of the router to utilize the available wiring channels on the substrate. The efficiency of a router is commensurate with the techniques employed for minimizing routing length and the number of vias, both of which help in reducing the number of signal layers. Typical routing efficiencies range from 40 to 70 percent. In general, the higher the routing efficiency, the lower the number of signal layers.

3.2 Density Approach

The density approach was initially developed for estimating the number of layers in printed wiring boards. A reference device is selected, and the number of signal layers is estimated as a function of the routing area available per reference device. This section demonstrates the application of the density approach to printed wiring boards, using the equivalent integrated circuit count method.

3.2.1 General estimation process

The number of signal layers is estimated as a function of the substrate density:

$$N_{layer} = f\left(\frac{1}{D_n}\right) \tag{3.4}$$

where N_{layer} is the number of signal layers and D_n is the substrate density. Substrate density is defined as the routable substrate area available per reference device:

$$D_n = \frac{A_r}{N_{ref}} \tag{3.5}$$

where A_r is the routable substrate area and N_{ref} is the total number of reference devices present on a substrate. The exact relationship between the number of signal layers and the density depends upon the type of the substrate and the reference device configuration. However, the number of signal layers is inversely proportional to substrate density. As the density increases, the routing area available for a reference device increases, and the required number of signal layers decreases.

A single reference device is selected for each type of substrate and is used to compare the wiring demand of the device configurations mounted on the substrates. The correlation factor of the ith device is defined as the equivalent number of reference devices that provide the same wiring demand as the ith device.

$$F_i = \frac{\text{Wiring demand of } i\text{th device}}{\text{Wiring demand of reference device}} \qquad (3.6)$$

where F_i is the correlation factor of the ith device with respect to the reference device. The total number of reference devices is the summation of the correlation factors of all the devices mounted on the substrate with respect to the reference device.

$$N_{ref} = \sum_{i=1}^{N_d} F_i \qquad (3.7)$$

Thus, the selection of the reference device configuration is critical to the estimation process because the device must facilitate easy comparison of the wiring demand with other device configurations.

3.2.2   Equivalent integrated circuit count method

The equivalent integrated circuit count method [PAD88, HAR89] is widely used in the industry for layer estimation in printed wiring boards. In this technique, a 14-pin through-hole DIP is selected as the reference device and defined as a unit Equivalent Integrated Count (EIC). All the devices mounted on the printed wiring board are then compared to the 14-pin DIP to determine the correlation factors.

One of the most popular methods of determining the correlation factors for DIPs, is to calculate the ratio of the number of leads between the DIP under consideration and the 14-pin DIP:

$$F_i = \frac{N_{li}}{14} \quad \{\text{i: DIP}\} \qquad (3.8)$$

where N_{li} is the number of leads of the ith DIP package. The number of interconnections made by the ith DIP and the footprint area of the ith DIP are assumed to be proportional to the number of device leads. For example, the footprint area of a 14-pin DIP is the summation of the pad areas of the fourteen leads in all the board layers. The footprint area of a 64-pin DIP is the summation of the pad areas of the sixty four leads in all the board layers. Thus, the correlation factor of the ith DIP package can be determined as a ratio of the number of leads of the ith device to the 14-pin DIP. Table 3.2 provides the correlation values for a few device configurations [HAR89].

Table 3.2. Correlation factors of selected device configurations

Device configuration	Correlation factor
Resistors/Capacitors	0.5
14-16 pin-DIP	1.0
64 pin-DIP	4.5
68 pin-PGA	4.5
149 pin-PGA	12
Single inline package	0.5
Relay	1.0
Edge card connector	Number of leads/64
High density connector	Number of leads/14

The correlation factors are used for calculating the total number of reference devices and the board density. The number of signal layers is estimated as a function of the board density. Example layer recommendations by Pads-Pcb [PAD88] and Hardenberg [HAD89] for different density values are listed in Table 3.3. A density of 1.0 means that 1.0 sq.in. of routable area is available on the board for routing a single 14-pin DIP. The typical package area of a 14-pin DIP is 0.32 sq.in. Table 3.3 suggests that approximately three times the package area is needed to rout a 14-pin DIP in two layers. The density ranges for various number of layers were determined by trial and error, and differ from one routing package to another.

Table 3.3. Layer recommendation for printed wiring boards by EIC method

Density (sq.in./14 pin DIP)	Pads-Pcb recommendation	Hardenberg's recommendation
above 1.0	2	2 with 20 percent difficulty
0.80 - 1.0	2	4
0.60 - 0.8	4	4+
0.42 - 0.6	6	N/A
0.35 - 0.42	8	N/A
0.20 - 0.35	10	N/A
0.0 - 0.2	10+	N/A

The inverse relationship of the number of signal layers to density makes the EIC estimation model sensitive for densities less than 0.2. Figure 3.2 shows graphically the relationship between density and the number of signal layers for the EIC model

[PAD88]. The graph shows that the estimation process is accurate for a density range of 0.2 to 1.0 (number of layers less than ten). There is a sharp increase in the slope of the curve for density less than 0.2 indicating that the estimation process is very sensitive for substrates with more than ten layers. Thus, the EIC approach does not provide an accurate estimate for substrates with more than ten layers.

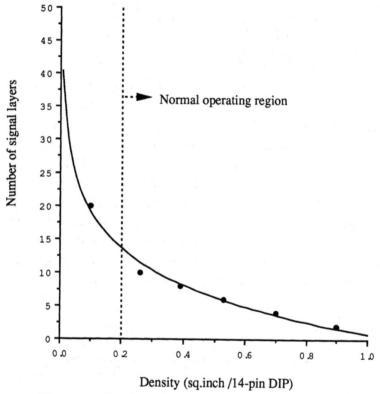

Density (sq.inch /14-pin DIP)

Figure 3.2. Number of signal layers vs. density: EIC model

Example: Consider a Pads-pcb board [MAR88] of 108 sq.in. with 241 through-hole components. The routable board area was found to be 103.8 sq.in. Table 3.4 presents the different device configurations and the number of devices in each device configuration. The correlation factors for the device configurations are obtained from Table 3.2. The total number of reference devices is determined

from Equation 3.7 as 232. The board density is found to be 0.45 sq.in./14-pin
DIP, using Equation 3.5. The number of signal layers required for the printed
wiring board is estimated from Table 3.3 as six. The actual number of signal
layers was found to be four [PAD88].

Table 3.4. Reference device calculation: example application

Device configuration	Number of devices	Correlation factor	Number of reference devices
DIP-14	4	1	4
DIP-16	27	1	27
DIP-18	37	1.2	44.4
DIP-20	28	1.2	33.6
DIP-28	2	1.8	3.6
DIP-40	1	2.5	2.5
DIP-62	9	4.5	40.5
PGA-68	2	4.5	9
PGA-140	1	12.0	12.0
HCONN-62	1	4.3	4.3
SIP	13	0.5	6.5
AXL	116	0.5	58

◇

3.2.3 Comments on the density approach

The effects of two important factors, conductor dimensions and percentage of
signal interconnections, are excluded by the estimation model. For example, two
substrates with the same set of devices but different conductor dimensions (trace
width of 12 mils and 4 mils) are recommended to have the same number of layers.
This result is not accurate because the number of layers normally decreases with
the decrease in conductor dimensions. Similarly, two substrates with the same set
of devices but different percentage of signal interconnections (30 percent and 70
percent) are recommended to have the same number of signal layers. However,
the number of signal layers normally increases with the increase in the percentage
of signal interconnections.

In the same substrate, the selection of a reference device also limits the determination of the correlation factors to certain device configurations. For example,
the 14-pin DIP serves as a suitable reference in printed wiring boards where the
components are mostly of through-hole configuration. In the case of surface mount
devices, the 14-pin device does not serve as a good reference for comparing the

wiring demands, because the footprint area in a surface mount device is restricted only to the mounting layer and does not extend to the inner layers. Thus, the computation of the correlation values as a ratio of the number of leads to the 14-pin DIP does not accurately reflect the ratio of the routing area demands. With the present shift of technology from through-hole to surface mount devices, a surface mount configuration will be an appropriate reference for comparing the wiring demands. However a better option is to estimate the number of signal layers independent of a reference device.

Furthermore, the selection of a reference device for each type of substrate also makes the estimation process substrate specific. For example, the 14-pin DIP package is generally restricted only to printed wiring boards and not commonly used in hybrid or multichip modules. Thus, the relationship between density and the number of layers used in the EIC method is applicable only for printed wiring boards and special relationships must be defined for hybrids and multichip modules.

3.3   Connectivity Approach

The concept of connectivity was first introduced by Messner [MES87] for comparing the wiring capabilities of different substrates, such as hybrids, multichip modules and printed wiring boards. This section describes the formulation of the signal layer estimation model using the connectivity approach. Unlike the EIC model developed using the density approach, this model can be applied to a variety of substrates because the wiring demand is represented independently of the substrate type and reference device.

Figure 3.3 shows the configuration of a typical substrate with devices of varied sizes and mounting styles. The number of signal layers is estimated by assuming a uniform wire distribution and by evaluating the wiring demand of all the devices. The wiring demand of a device is expressed in terms of connectivity, defined as the routing length per unit substrate area. The number of signal layers required is expressed as the ratio of the average demanded to permitted connectivity of the devices.

$$N_{layer} = \frac{1}{N_d} \sum_{i=1}^{N_d} \left(\frac{C_{di}}{C_{pi}} \right) = ave \left(\frac{C_{di}}{C_{pi}} \right) \tag{3.9}$$

where C_{di} is the demanded connectivity of the ith device, C_{pi} is the permitted

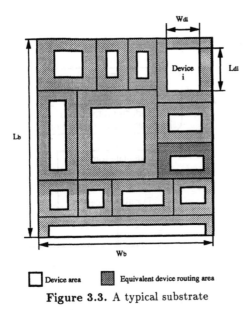

Figure 3.3. A typical substrate

connectivity of the ith device, and N_d is the total number of devices. Figure 3.4 presents a brief overview of the overall estimation process.

The inputs to the estimation process were listed in Table 3.1, and the routable substrate area was explained in Section 3.1. Permitted connectivity is the routing length permitted by the device configuration and conductor dimensions, as described in Section 3.3.1. Models are formulated for the permitted connectivity of through-hole and surface-mount devices; the behavior of the models is discussed by analyzing the connectivity values of commonly used devices.

Demanded connectivity is the routing length demanded by a device and consists of two types, interconnect connectivity and access connectivity, as shown in Figure 3.4. Interconnect connectivity is the routing length demanded by a device for making interconnections with other devices. Section 3.3.2.1 describes the formulation of interconnect connectivity as a function of the routable substrate area, percentage of signal interconnections, and average routing length per interconnection. A new technique, which uses an uniform pin distribution, is recommended for predicting the average routing length per interconnection, and is compared with existing routing length prediction techniques. Section 3.3.2.2 describes the formulation of access connectivity, which is the additional routing length required by the device configuration for accessing pins located in the package interior.

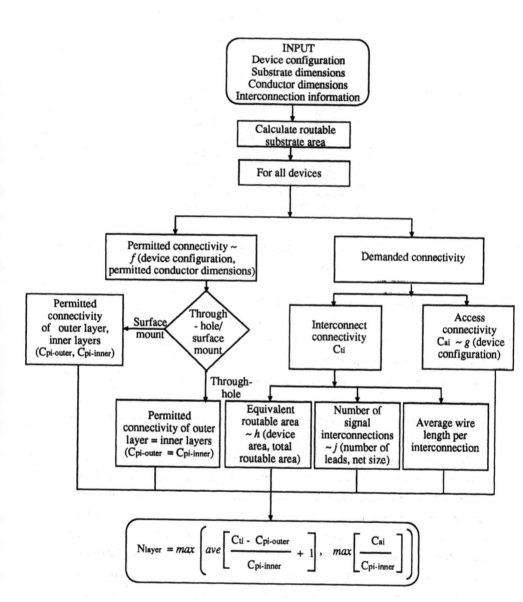

Figure 3.4. Flowchart for layer estimation using connectivity approach

The permitted and demanded connectivity models are combined to develop a generic layer estimation model in Section 3.3.3. Modifications of the model for special cases, such as through-hole printed wiring boards and multichip modules, are also discussed. The section concludes with a demonstration of the layer estimation process through an example application.

3.3.1 Permitted connectivity

The permitted connectivity of a device is defined as the routing length per unit area allowed by the device configuration and the conductor dimensions. This section describes the formulation of permitted connectivity of through-hole and surface-mount devices.

In a typical routing substrate, signal interconnections between devices are made by traces that run between signal via-pads and device mounting-pads. Signal via-pads are used to connect signal traces located on different layers. Device mounting-pads are used to support the device leads and might be located on all the substrate layers, or only on the substrate surface, depending upon the device configuration. The space between two adjacent via-pads or mounting-pads is called the channel width.

The permitted connectivity of a device is governed by the number of traces

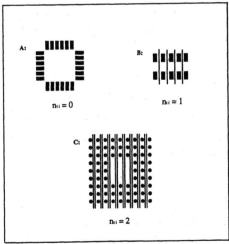

Figure 3.5. Permitted connectivity of selected device configurations

that are permitted through the channel width. The number of traces permitted in a channel depends upon the device configuration, trace width, trace spacing, and pad dimensions. Figure 3.5 shows three different device configurations that permit different number of traces per channel due to differences in their lead pitch and mounting-pad dimensions.

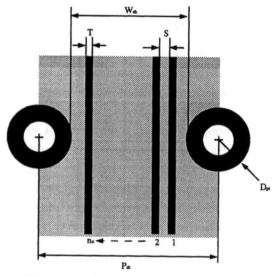

Figure 3.6. Typical conductor configuration

Figure 3.6 shows the conductor configuration between two adjacent mounting-pads. The number of traces permitted through the channel width of the ith device, n_{ti}, is given by:

$$n_{ti} = int\left(\frac{P_{di} - D_{pi} - S}{T + S}\right) \tag{3.10}$$

where T is the trace width, S is the trace spacing, D_{pi} is the mounting-pad width/diameter, P_{di} is the lead pitch of the ith device, and $int\,(x)$ is the integer part of the real number, x. The total trace length, L_{pi}, permitted in the substrate area, P_{di}^2, between two adjacent pads is given by:

$$L_{pi} = n_{ti}\, P_{di} \tag{3.11}$$

Thus, the permitted connectivity of a device, C_{pi-f}, is the ratio of the trace length permitted by the device footprint to the substrate area.

$$C_{pi-f} = \frac{L_{pi}}{P_{di}^2} = \frac{n_{ti}}{P_{di}} \tag{3.12}$$

• Permitted connectivity of surface-mount devices

In some cases, the permitted connectivity of a device is not only dependent on the device footprint, but also on the signal via dimensions. For example, in surface-mount devices, the device leads are mounted only on the substrate surface and the device footprints are thus located only on the mounting/outer layer. Connections to the inner signal layers are established by signal vias. Thus, the permitted connectivity of the outer layer, $C_{pi-outer}$, is governed by the device footprint and the permitted connectivity of the inner layers, $C_{pi-inner}$, is governed by the signal via dimensions. Thus,

$$C_{pi-outer} = C_{pi-f} \quad \{i: \text{surface mount device}\} \qquad (3.13)$$

$$C_{pi-inner} = \frac{n_{tsv}}{P_{sv}} \quad \{i: \text{surface mount device}\} \qquad (3.14)$$

where P_{sv} is the signal via-pitch and n_{tsv} is the number of traces permitted between the signal via-pads and given by:

$$n_{tsv} = \frac{P_{sv} - D_{psv} - S}{T + S} \qquad (3.15)$$

where D_{psv} is the signal via-pad diameter. As the number of signal layers increases in surface-mount substrates, the effect of the permitted connectivity of inner layers dominates over the effect of the permitted connectivity of the outer layer. For example, let the permitted connectivity of the outer layer be 10 in./sq.in. and the permitted connectivity of the inner layers be 60 in./sq.in. If the substrate contains two layers, the average permitted connectivity of the substrate, $C_{p(2-layers)}$ = (10 + 60)/2 = 35 in./sq.in., and if the substrate contains ten layers, the average permitted connectivity of the substrate, $C_{p(10-layers)}$ = (10 + 9 x 60)/10 = 55 in./sq.in. Thus, the permitted connectivity of the outer layer is critical in substrates with a lower number of signal layers.

• Permitted connectivity of through-hole devices

In through-hole devices, the device leads pass through all the layers of the substrate and the permitted connectivity of both the outer and inner layers is governed by the through-hole device footprint.

$$C_{pi-outer} = C_{pi-inner} = C_{pi-f} \quad \{i: \text{through hole device}\} \qquad (3.16)$$

Figure 3.7 shows the permitted connectivity of through-hole and surface-mount devices as a function of the device lead pitch and number of traces per channel.

The figure shows that the permitted connectivity of a device increases with an increase in the lead pitch. As the lead pitch increases, the channel width increases, thereby increasing the number of traces permitted through the channel. Thus, the permitted connectivity is directly proportional to the lead pitch. In contrast, as the trace width and spacing decrease, more traces are permitted through the channel and the permitted connectivity increases. For example, for a lead pitch of 0.100 in., the permitted connectivity with 0.012 in. trace width is 15 in./sq.in., while the permitted connectivity with 0.004 in. is 60 in./sq.in.

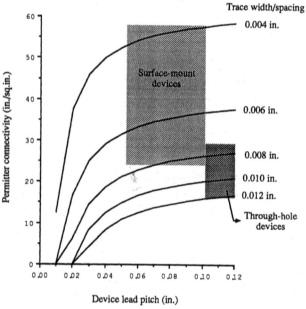

Figure 3.7. Permitted connectivity

The typical lead pitch in through-hole devices is 0.100-in., and the lead-pad diameters range from 0.040-in. to 0.060-in. The commonly observed permitted connectivity values of through-hole devices range from 10 in./sq.in. to 25 in./sq.in. Typical surface-mount device lead pitches range from 0.050-in. to 0.100-in., mounting-pad widths range from 0.020-in. to 0.030-in., and signal via-pad diameters range from 0.020-in. to 0.040-in. The via-pad diameters and lead pitches are smaller in surface-mount devices compared with the through-hole devices because the signal vias are used only for establishing electrical connections, and not for supporting the leads. In general, the permitted connectivity of the outer layer ranges from 0 to 40 in./sq.in. and the permitted connectivity of the inner layers ranges from 20 to 60 in./sq.in. The decrease in the permitted connectivity on the outer layer is due to the reduced lead pitch in surface-mount devices.

In summary, the permitted connectivity of surface-mount devices is governed by both the device footprint and the signal via dimensions. The permitted connectivity of through-hole devices is governed only by the device footprint. Though the permitted connectivity of the outer layer is not significant in surface-mount devices, the permitted connectivity of the inner layers is almost 1.5 to 2 times the permitted connectivity of through-hole devices. This increased permitted connectivity of the inner layers, along with the reduced size of the devices, is responsible for the high effective permitted connectivity in surface-mount substrates.

3.3.2 Demanded connectivity

In this section, the focus is on the determination of the routing length demanded by devices for making signal interconnections. Demanded connectivity is defined as the wiring length per unit area required by a device, and is categorized as either interconnect connectivity and access connectivity. In general, the demanded connectivity of a device is governed by the interconnect connectivity, and the effect of access connectivity is usually omitted in layer estimation models [MES87, COO88, MAR89]. However, the significance of access connectivity is critical in devices with a high pin density and area-array packages. Devices with low lead pitches require a large fanout area for routing and area array packages require traces for accessing leads located in the package interior. The access connectivity accounts for the increased wiring demand imposed due to the device configuration, and is thus independent of the available routable substrate area. The interconnect connectivity is formulated as a function of the routable substrate area, the number of signal interconnections, and the average routing length per interconnection. The access connectivity is formulated as a function of the device configuration.

3.3.2.1 Interconnect connectivity

The interconnect connectivity of a device is defined as the routing length per unit area required by the device for making signal interconnections with other devices.

The interconnect connectivity is a function of the total number of signal interconnections made by the device, the average routing length per interconnection, and the routable substrate area available for the device.

The total routing length, L_{di}, demanded by a device for interconnections is given by:

$$L_{di} = L_{avg} \ N_{wi} \qquad (3.17)$$

where N_{wi} is the number of signal interconnections made by the ith device, and L_{avg} is the average routing length per interconnection. If the equivalent routable area available on the substrate for routing the ith device is denoted as A_{ei}, the interconnect connectivity, C_{ti}, is given by the ratio of the demanded routing length to the equivalent routable area:

$$C_{ti} = \frac{L_{di}}{A_{ei}} = \frac{L_{avg} \ N_{wi}}{A_{ei}} \qquad (3.18)$$

Perhaps the most important part of the layer estimation process is the determination of the interconnect connectivity. Several attempts have been made in the past to formulate interconnect connectivity by using Equation 3.18, and a few approaches are discussed here. The differences in these approaches arise from the assumption of different values for the average routing length number of signal interconnections, and equivalent routable area. Welterlen extended Seraphim's [SER78] routing length equation to obtain a normalized interconnect connectivity for square devices:

$$C_{ti(Welterlen)} = \frac{K_w \ N_{li}}{\frac{N_{li} P_{di}}{4} + 5 \, P_{di}} \ (in./sq.in) \qquad (3.19)$$

where N_{li} is the number of device leads, P_{di} is the lead pitch of the device in inches, and K_w is a fudge factor defined by Welterlen as 2.25. Equation 3.19 is based on the assumption that the spacing between all devices is five times the lead pitch of the device ($5 \, P_{di}$), independent of the available routable substrate area. This inter-device spacing is too small for devices with lead pitches less than 25 mils and results in high wiring demands. For example, a lead pitch of 20 mils results in an inter-device spacing of only 100 mils.

Messner [MES87] defines the interconnect connectivity of a device as:

$$C_{ti(Messner)} = \frac{K_m \ N_{li}}{S_d} (in./sq.in.) \qquad (3.20)$$

where S_d is the pitch between two devices and K_m is Messner's constant, which equals 1.125. However, the pitch between two devices can be obtained only after device placement. Bogatin [BOG90] defines the interconnect connectivity as:

$$C_{ti_{(Bogatin)}} = \frac{K_b N_{li}}{A_{di}} (in./sq.in.) \qquad (3.21)$$

where A_{di} is the device area and K_b is Bogatin's constant, which equals 1.5.

Coombs [COO88] assumes that each device pin requires an average routing length of 2.25 in. independent of the substrate size or device configuration:

$$C_{ti_{(Coombs)}} = \frac{K_c N_{li}}{A_{di}} (in./sq.in.) \qquad (3.22)$$

where K_c is Coomb's constant, which equals 2.25. Both Coomb's and Bogatin's methods assume that the equivalent routable area available for a device is restricted only to the device area. For example, the shaded regions of the substrate shown in Figure 3.3 are not used for routing, and hence result in high interconnect connectivity values.

All the models assume that a device is connected only to an adjacent device. This results in low estimates of the average routing length for substrates where connections are made between non-adjacent devices on the substrate. The models also assume a uniform array of a single device on the substrate, a configuration that may be common for multichip module substrates, but not for most hybrids or printed wiring boards. The subsequent sections describe the present approach for determining equivalent routable area, number of signal interconnections, and average routing length, which will later be substituted in Equation 3.18 to determine interconnect connectivity.

• Equivalent routable area

The equivalent routable area of a device is defined as the substrate area available for routing the device. Figure 3.3 shows a substrate with device areas and their equivalent routing areas. Assuming that the total routable substrate area, A_r, is distributed among the devices as equivalent routable areas:

$$A_r = \sum_{i=1}^{N_d} A_{ei} \qquad (3.23)$$

where A_{ei} is the equivalent area of the ith device. The equivalent area of a device can be expressed as a fraction of the total routable area, A_r, and proportional to the device area, A_{di} as:

$$A_{ei} = \frac{A_{di} A_r}{A_d} \tag{3.24}$$

where A_d is the total device area obtained from Equation 3.2. The ratio of the total device area to routable substrate area is defined as the area utilization factor, A_{uf}:

$$A_{uf} = \frac{A_d}{A_r} \tag{3.25}$$

Substituting for A_{uf} in Equation 3.24, the equivalent routing area for a device can be obtained as:

$$A_{ei} = \frac{A_{di}}{A_{uf}} \tag{3.26}$$

For a given set of devices and routable substrate area, the area utilization factor of the substrate remains constant. The area utilization factor also represents the percent routable substrate area occupied by devices. The value of A_{uf} lies between 0 and 1. An A_{uf} value greater than 1.0 implies that the total device area exceeds the substrate routable area, and hence devices cannot be placed on the substrate.

• Number of signal interconnections

The number of signal interconnections made by a device may be provided as an input with the interconnection information or can be estimated as a function of the net size and number of device leads. The number of signal interconnections made by a device, N_{wi}, can be expressed as:

$$N_{wi} = \frac{N_{li}(n - 1)}{n} \tag{3.27}$$

where N_{li} is the number of device leads and n is the net size. Table 3.5 shows that variation of the number of signal interconnections for various net sizes is a function of the number of leads.

Table 3.5. Relationship between net size and number of interconnections

Net size	Number of interconnections
2	$0.50\ N_{li}$
3	$0.66\ N_{li}$
4	$0.75\ N_{li}$
5	$0.8\ N_{li}$

In general, as the net size increases, the number of signal interconnections also increases. Typical net sizes observed in most electronic circuits are two or three [MES87, TOL91]. Thus, the number of signal interconnections normally observed is 50 to 70 percent of the number of device pins. If the average net size of a circuit is unknown, then a default net-size of two may be assumed.

The ratio $(n-1)/n$ is also called the interconnection-to-pin ratio and represents the percentage of device pins used for making signal connections. Thus, if the total number of signal interconnections is provided rather than the average net size, the interconnection-to-pin ratio is calculated as the ratio of the number of signal interconnections to the total number of device pins. In summary, the number of signal interconnections made by a device is:

$$N_{wi} = I_{pr} N_{li} \qquad (3.28)$$

where I_{pr} is the interconnection-to-pin ratio given by:

$$I_{pr} = \begin{cases} (n-1)/n & \text{if } n \text{ is given} \\ N_w / \sum_{i=1}^{N_d} N_{li} & \text{if } N_w \text{ is given} \\ 0.5 & \text{default} \end{cases} \qquad (3.29)$$

where N_w is the total number of signal interconnections.

• Average routing length per interconnection

The challenging and critical part of the layer estimation process is predicting the average routing length per interconnection, because very little information is available about the circuit topology prior to device placement. The starting and ending positions of the interconnections are unknown, and the average routing length is estimated by assuming a uniform wire distribution as a function of the number of device leads, the routable substrate area and the net size. Most of the research work in this area has been focussed on determining the average routing length [SER78, DON79, TUM89] on a square substrate. However, Donath [DON79] has noted that the average routing length on a square and rectangular substrate of the same area are significantly different.

The next section describes the uniform pin distribution technique for predicting the average routing length as a function of the substrate area, aspect ratio and the total number of device pins. The subsequent sections discuss other estimation techniques and compare the accuracy of this estimation technique with existing techniques.

• Uniform pin distribution technique for average routing length estimation

The devices are assumed to be uniformly distributed on a substrate of length L_b, width W_b and routable area of A_r as shown in Figure 3.8. It is also assumed that the routing length demand of the diagonal pins is representative of the routing length demand of all the device pins.

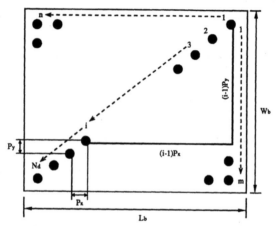

Figure 3.8. Substrate configuration: Uniform pin distribution technique

If there are X pins along the L_b direction and Y pins along W_b direction, then,

$$N_l = XY \qquad (3.30)$$

and

$$\frac{Y}{X} = \frac{W_b}{L_b} \qquad (3.31)$$

where N_l is the total number of device pins. From Equations 3.30 and 3.31, the number of pins along the diagonal, N_{dg}, is:

$$N_{dg} = \sqrt{\frac{N_l(L_b^2 + W_b^2)}{2L_bW_b}} \qquad (3.32)$$

Let P_x be the pitch between two adjacent N_{dg} pins in the L_b direction, and P_y be the pitch between two adjacent N_{dg} pins in the W_b direction. If $P_{N_{dg}}$ is the total Manhattan routing length between two adjacent diagonal pins, then it can be expressed as:

$$P_x = \frac{L_b}{N_{dg}} = \sqrt{\frac{2L_b^3 W_b}{N_l(L_b^2 + W_b^2)}} \qquad (3.33)$$

$$P_y = \frac{W_b}{N_{dg}} = \sqrt{\frac{2W_b^3 L_b}{N_l(L_b^2 + W_b^2)}} \qquad (3.34)$$

Now the problem is reduced to estimating the routing length between N_{dg} points along the diagonal with a routing length of $P_{N_{dg}}$ between two adjacent points. Figure 3.8 shows that pin 1 can be connected either to pin 2 or pin 3 or....pin N_{dg}. Correspondingly, the routing length for each of the connections is $P_{N_{dg}}$, $2P_{N_{dg}}$, $3P_{N_{dg}}$, ..., and $(N_{dg}\text{-}1)P_{N_{dg}}$, assuming that all the connections are of two-terminal nets. Since the total number of possible interconnections is $(N_{dg} - 1)$, the probability of a single connection is $1/(N_{dg} - 1)$. Thus, the average routing length for pin 1 is given by:

$$A_{wl1} = \frac{P_{N_{dg}} + 2P_{N_{dg}} + 3P_{N_{dg}} + ... + (N_{dg} - 1)P_{N_{dg}}}{N_{dg} - 1} \qquad (3.35)$$

Similarly, the average routing length of N_{dg-1} pins (A_{wl1}, A_{wl2} ..., A_{wlNdg}) can be expressed as:

$$
\begin{bmatrix}
0 & 1 & 2 & 3 & \cdots & N_{dg} - 1 \\
1 & 0 & 1 & 2 & \cdots & N_{dg} - 2 \\
2 & 1 & 0 & 1 & \cdots & N_{dg} - 3 \\
\cdot & \cdot & \cdot & \cdot & \cdots & \cdot \\
\cdot & \cdot & \cdot & \cdot & \cdots & \cdot \\
\cdot & \cdot & \cdot & \cdot & \cdots & \cdot \\
N_{dg} - 2 & \cdot & \cdot & \cdot & \cdots & 1 \\
N_{dg} - 1 & \cdot & \cdot & \cdot & \cdots & 0
\end{bmatrix}
\begin{bmatrix}
1 \\ 1 \\ 1 \\ \cdot \\ \cdot \\ \cdot \\ 1 \\ 1
\end{bmatrix}
\frac{P_{N_{dg}}}{N_{dg} - 1} =
\begin{bmatrix}
A_{wl1} \\ A_{wl2} \\ A_{wl3} \\ \cdot \\ \cdot \\ \cdot \\ A_{wlN_{dg}-1} \\ A_{wlN_{dg}}
\end{bmatrix}
\qquad (3.36)
$$

Assuming that the average routing length of the diagonal pins gives the average routing length per interconnection for the substrate:

$$L_{avg} = \frac{A_{wl1} + A_{wl2} + \cdots + A_{wlN_{dg}}}{N_{dg}} \qquad (3.37)$$

Solving Equations 3.36 and 3.37,

$$L_{avg} = \frac{P_{N_{dg}}(N_{dg} + 1)}{6} \qquad (3.38)$$

Substituting the values of $P_{N_{dg}}$ and N_{dg} in Equation 3.38:

$$L_{avg} = \frac{L_b + W_b}{6}[1 + \sqrt{\frac{2L_b W_b}{N_l(L_b^2 + W_b^2)}}] \qquad (3.39)$$

The aspect ratio of the substrate is defined as the ratio of the width to the length of the substrate (W_b/L_b). The value of the aspect ratio, A_s, ranges from 0 to 1.0. The routable substrate area, A_r, can be approximated as substrate area $(L_b W_b)$. Extending Equation 3.39 for n terminal nets and substituting for A_s and A_r:

$$L_{avg(upd)} = \frac{\sqrt{A_r}(A_s + 1)}{6(n-1)\sqrt{A_s}}[1 + \sqrt{\frac{2A_s}{N_l(A_s^2 + 1)}}] \qquad (3.40)$$

For a square substrate, the aspect ratio is 1.0 and the average routing length per interconnection is reduced to:

$$L_{avg(upd)} = \frac{0.34(1 + \sqrt{N_l})}{n-1}\sqrt{\frac{A_r}{N_l}} \qquad (3.41)$$

Equation 3.41 shows that the average routing length per interconnection is proportional to the square root of the routable substrate area. With the increase in routable substrate area, the spacing between two devices increases, resulting in an increase in the average routing length. This result is expected because of the assumption that the devices are uniformly spread on the substrate.

The average routing length also has an inverse relationship with the total number of device pins and net size. When the total number of device pins on a substrate increases, the pins are located closer to one another and the average routing length between the pins decreases. Similarly, when the net size increases, more terminals are available in a single net, and the routing length between two adjacent terminals decreases. The minimum average routing length is obtained for a square substrate $(A_s = 1.0)$, because the diagonal length of a square is always smaller than the diagonal length of a rectangle of the same area. The average routing length increases with the decrease in the aspect ratio $(W_b < L_b)$. In general, the effect of the aspect ratio on the average routing length is not significant when the aspect ratio is between 0.5 and 1.0. However, for aspect ratios less than 0.5, a significant increase in the average routing length occurs, due to the non-uniform distribution of routing length in the horizontal and vertical wiring planes in substrates with high aspect ratios.

• Seraphim's technique for average routing length estimation

Seraphim [SER78] derived the average routing length per interconnection for a uniform distribution of Nd devices (of same package style and number of pins) spaced with a uniform pitch, S_d, (center to center spacing between two adjacent devices) as:

$$L_{avg(ser)} = 1.5\,S_d \qquad (3.42)$$

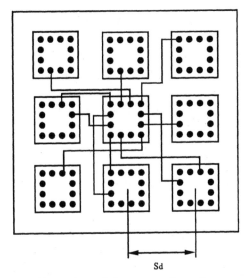

S_d

Figure 3.9. Substrate configuration

In Figure 3.9 [SER78, TUM89], the centrally located device is connected to the neighboring device with an average Manhattan distance S_d, or to the next device beyond, for an average Manhattan distance of $2\,S_d$. Averaging the two distances gives the $1.5\,S_d$ of Equation 3.42. The average pitch, S_{d-ave}, between the devices can be expressed in terms of the total routable area and the number of devices as:

$$S_{d-ave} = \sqrt{\frac{A_r}{N_d}} \qquad (3.43)$$

With devices of varied sizes, S_{d-ave} computed from Equation 3.43, is a poor assumption, because significant differences exist between the device center-center pitches. Substituting for S_d with S_{d-ave}:

$$L_{avg(ser)} = 1.5\,\frac{A_r}{N_d} \qquad (3.44)$$

• Tummala's technique for average routing length estimation

   Tummala [TUM89] assumed a similar device configuration as Seraphim and obtained a logarithmic distribution for the average routing length by statistical analysis as:

$$L_{avg(tum)} \approx 0.77 \, N_d^{0.245} \, S_d \tag{3.45}$$

Substituting for S_d with S_{d-ave} in Equation 3.45:

$$L_{avg(tum)} \approx 0.77 \, N_d^{-0.255} \, A_r^{0.5} \tag{3.46}$$

• Donath's technique for average routing length esitmation

   Donath [DON79] estimated the average routing length in terms of routable area, number of devices and the electrical characteristics of the circuit as:

$$L_{avg(don)} = \frac{2A_r^{0.5}}{9N_d^{-0.5}} \left(7\frac{N_d^{\beta-0.5}-1}{4^{\beta-0.5}-1} - \frac{1-N_d^{\beta-0.75}}{1-4^{\beta-0.75}}\right)\frac{1-4^{\beta-1}}{1-N_d^{\beta-1}} \tag{3.47}$$

where β is the Rent's constant. Bokaglu [Bok91] recommends $\beta = 0.25$ for system level boards and $\beta = 0.63$ for modules.

• Comparison of average routing length estimation techniques

Table 3.6 shows the estimated routing lengths from the different techniques and comparison with actual values for five substrates [MUR90, NAK90]. Rent's constant is assumed to be 0.25 for Donath's [DON79] estimation, because most of the test cases are printed circuit boards.

Table 3.6.   Comparison of average routing length estimation techniques

Test #	Substrate area	# of pins	Net size	Average routing length (in.)					Error (in.)			
				ser	tum	don	upd	actual	ser	tum	don	upd
1*	276.5	14537	3	0.80	2.22	2.57	2.80	1.92	-1.12	0.30	0.65	0.53
2*	276.5	16960	3	0.78	2.19	1.32	2.69	2.61	-0.83	-0.42	-1.29	0.08
3**	407.4	60480	3	1.65	3.52	2.63	3.38	3.91	-2.26	-0.39	1.28	-0.53
4*	330	16174	2	0.92	1.54	1.56	6.10	4.53	-3.61	-4.56	-4.54	1.57
5*	351.5	19452	2	1.10	2.26	1.84	6.29	6.39	-5.29	-4.03	-4.45	-0.10

Reference: * [MUR90]   ** [NAK90]

Figure 3.10 provides a graphical comparison of the average routing length estimates. The uniform pin distribution technique provided the least error among the estimation techniques. Seraphim's [SER78] technique gives low estimates, because of the assumption that a device is connected only to its adjacent neighbors. Donath's [DON79] and Tummala's [TUM89] estimates lie between Seraphim's and the uniform pin distribution technique. In Donath's technique, the estimation process is highly sensitive to the Rent's constant. The present layer estimation process uses the uniform pin distribution technique.

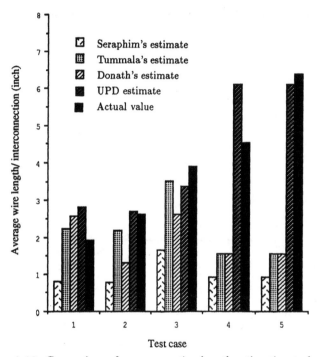

Figure 3.10. Comparison of average routing length estimation techniques

• Interconnect connectivity model

The number of signal interconnections, Nwi, from Equation 3.28 and the equivalent routable area, A_{ei}, from Equation 3.26 are substituted in Equation 3.18 to obtain the interconnect connectivity of a device as:

$$C_{ti} = \frac{L_{avg}\, I_{pr}\, A_{uf}\, N_{li}}{A_{di}} \tag{3.48}$$

In Equation 3.48, the interconnection-to-pin ratio, average routing length per interconnection and area utilization factor remain constant for a substrate with a selected set of devices. Equation 3.48 shows that the interconnect connectivity of a device is directly proportional to the device pin density. Device pin density is defined as the number of device pins per unit device area.

$$C_{ti} = K \frac{N_{li}}{A_{di}} \tag{3.49}$$

where constant $K = L_{avg} I_{pr} A_{uf}$ and N_{li}/A_{di} is the device pin density. Devices with high pin density demand more trace length in the given routable area, and hence a higher interconnect connectivity.

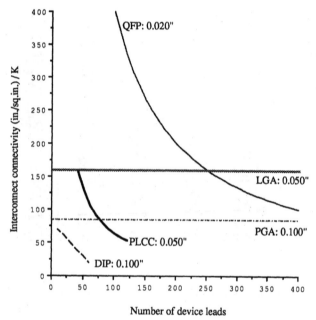

Figure 3.11. Interconnect connectivity of commonly used packages

Figure 3.11 shows the interconnect connectivity of some commonly used packages as a function of lead pitch and lead count. Area array packages (PGAs and LGAs) with a high pin density are found to have a higher interconnect connectivity compared to four-sided and two-sided packages (PLCCs and DIPs), which have a lower pin density. The interconnect connectivity is also inversely proportional to the device lead pitch. A significant increase in interconnect connectivity is observed between the PLCC of 0.050-in. lead pitch, and the QFP of 0.020 in. lead pitch. A similar effect is also observed between area-array packages of different lead pitch. The value of the constant, K, depends on the average routing length and number of signal interconnections and is responsible for the difference in the interconnect connectivity of two substrates having the same set of devices but different routable areas. In summary, the critical factors that influence the interconnect connectivity of a device are the device pin density and the lead pitch.

3.3.2.2   Access connectivity

Access connectivity is defined as the routing length demanded by a device to access the device leads from the package periphery. When the DIP technology was widely used, routers did not have much difficulty in accessing the device leads because the leads were located at the periphery of the package. Furthermore, an adequate lead pitch of 0.100-in. was provided for the passage of traces. With the introduction of device configurations like pin-grid arrays and quad flat packs, which have either rows of leads or smaller device lead pitches, accessing the leads has become increasingly difficult. The routing area demanded by a device is no longer governed by interconnect connectivity alone, but also by access connectivity. Thus it is important to integrate the effect of access connectivity in the layer estimation model. In this section, the access connectivity model is formulated by examining the access connectivity of the DIP and PGA packages. The access connectivity of commonly used packages is also examined.

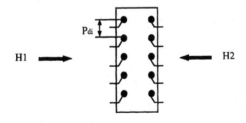

Figure 3.12. Access connectivity of a dual-inline package

Figures 3.12 and 3.13 demonstrate the access connectivity of a DIP and a PGA device. In general, every lead needs one trace for making a connection, and thus one trace is required to pass through the lead pitch. The DIP package has two rows of parallel leads, that can be accessed either by horizontal or vertical traces.

Assuming that the leads are accessed by horizontal traces, the leads on the right side are accessed by traces approaching from left to right (H2 in Figure 3.12). Similarly, the leads on the left side are accessed by traces approaching from right to left (H1 in Figure 3.12).

In effect, the number of accessing planes required for the DIP is one (either horizontal or vertical), because the rows of leads are parallel. By allowing one trace to pass through the device lead pitch P_{di}, (traces/channel $= 1$), the access connectivity can be calculated similarly to permitted connectivity as:

$$C_{ai} = \frac{n_{tai}}{P_{di}} = \frac{1}{P_{di}} \quad \{i: \text{DIP, Figure 3.12}\} \tag{3.50}$$

where n_{tai} is the number of traces per channel required for accessing and P_{di} is the device lead pitch.

The PGA in Figure 3.13 has four sides and five rows of pins from the periphery on a single side. Consider the pin I_{in}, located on a single side. If I_{in} is to be accessed from the periphery, five traces are required to pass through the lead pitch, P_{di}. Thus, the number of traces required for accessing is equal to the number of rows of pins from the periphery of the device. Assume that the pins located on the four sides are accessed from four directions (H1, H2, V1, V2 in Figure 3.13, with one pair of directions perpendicular to the other. Sides H1 and H2 are accessed using horizontal traces, and sides V1 and V2 are accessed using vertical traces. Thus, two accessing planes orthogonal to one another are required for the PGA. Extending Equation 3.50:

$$C_{ai} = A_p \frac{n_{tai}}{P_{di}} = 2\frac{R_i}{P_{di}} \quad \{i: \text{PGA, Figure 3.13}\} \tag{3.51}$$

where A_p is the number of accessing planes.

In general, the number of traces per channel required for accessing is equal to the number of rows from the periphery on a single side of the package and the number of accessing planes is governed by the shape of the device. Thus, the accessing connectivity, C_{ai}, is defined as:

$$C_{ai} = S_{fi} \frac{R_i}{P_{di}} \tag{3.52}$$

where S_{fi} is the shape factor of the ith device given by:

$$S_{fi} = \begin{cases} 1 & \text{two-sided device} \\ 2 & \text{four-sided device} \end{cases} \tag{3.53}$$

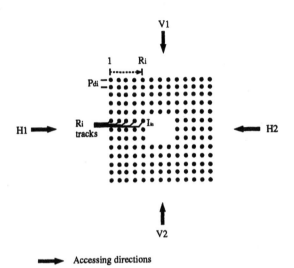

→ Accessing directions

Figure 3.13. Access connectivity of a pin grid array package

and R_i is the number of rows of pins from the periphery of the package on a single side. While the shape factor can be obtained from the device information, the number of rows in a device can be estimated by using the device dimensions L_{di}, W_{di} and P_{di}. Example S_{fi} and R_i values of commonly used device configurations are given Table 3.7.

Table 3.7. S_{fi} and R_i values for commonly used packages

Package style	S_{fi}	R_i
AXL-res/cap, DIP, SIP, SOIC, ECONN	1.0	1.0
PLCC, QFP	2.0	1.0
PGA, LGA	2.0	$k - \sqrt{k^2 - N_{li}^2}$ $(k^2 \geq N_{li})$, or 2 $(k^2 < N_{li})$
HCONN	1.0	W_{di}/P_{di}

where $k = L_{di}/P_{di}$

Figure 3.14 shows the accessing connectivity of common packages as a function of the lead count and lead pitch. In device configurations such as PGAs and LGAs, which allow rows of leads, the lead count strongly influences the access connectivity. In general, when the pin density increases, the number of rows of leads also increases, thereby increasing the access connectivity. The graph shows that the device lead pitch also plays an important role on the access connectivity. The device lead pitch determines the channel width available for the accessing traces. As the lead pitch decreases, the access connectivity increases. The access

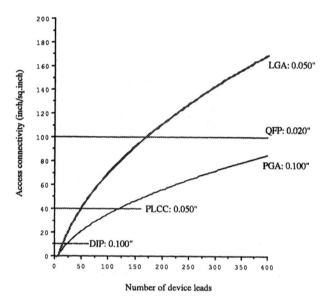

Figure 3.14. Access connectivity of commonly used packages

connectivity of a PLCC with 0.050 in. lead pitch is 40 in./sq.in., while the access connectivity of a QFP with 0.020-in. lead pitch is 100 in./sq.in., indicating that the QFP requires more routing area for fanout.

Among the various device configurations considered, the accessing connectivity of DIPs and PLCCs is not very significant. But the accessing connectivity of PGAs, LGAs and QFPs (with high lead counts and low lead pitches) plays a critical role in constraining the demanded connectivity of the device. Thus, the design team has to be cautious in the choice of certain device configurations. The use of a single 600-pin PGA on a predominantly DIP board might restrict the lower limit of the number of layers, independent of the demand of other devices.

The accessing connectivity of a device is not dependent on the routable substrate area, but only dependent on the device configuration and conductor dimensions. Thus, increasing the routing area will not reduce the access connectivity. For example, IBM's TCM modules (1800-pin PGA) needs a minimum of twenty layers for all the pins to be accessed independent of the substrate area on which the modules are mounted [BOG90].

3.3.3 Generic model for layer estimation

Equation 3.9 is expressed in terms of the permitted connectivity of outer and inner layers, the interconnect connectivity, and the access connectivity as:

$$N_{layer} = \text{ave}(\frac{C_{ti} - C_{pi-outer}}{C_{pi-inner}} + 1) \text{ or } \max(\frac{C_{ai}}{C_{pi-outer}}) \qquad (3.54)$$

In the case of surface-mount devices, the permitted connectivity of the inner and outer layers are different and can be obtained from Equations 3.13 and 3.14. Substituting for the interconnect and access connectivity from Equations 3.48 and 3.52, Equation 3.54 can be modified to obtain the final estimation model as:

$$N_{layer} = \frac{1}{R_e}\text{ave}(\frac{P_{sv}(P_{di}L_{avg}I_{pr}A_{uf}N_{li} - n_{ti}A_{di})}{n_{tsv}A_{di}P_{di}} + 1) \text{ or } \max(\frac{S_{fi}R_i}{P_{di}}) \qquad (3.55)$$

where R_e is the routing efficiency. Typical routing efficiencies range from 30 to 70 percent.

For the special case of through-hole printed circuit boards, the permitted connectivity of the outer/mounting layer is equal to the permitted connectivity of each of the inner layers, because the device leads pass through all the layers. Thus, the layer estimation model for through-hole printed wiring boards is

$$N_{layer} = \frac{P_{sv}}{R_e n_{tsv}} \text{ave}(\frac{L_{avg}I_{pr}A_{uf}N_{li}}{A_{di}}) \text{ or } max(\frac{S_{fi}R_i}{P_{di}}) \qquad (3.56)$$

In the case of multichip modules (MCM), devices of similar configuration are often distributed on the substrate. Thus, the demanded and permitted connectivities of a single device represent the demanded and permitted connectivities of the whole substrate. The area utilization factor, A_{uf}, for multichip modules is defined as:

$$A_{uf} = \frac{N_d A_{di}}{A_r} \qquad (3.57)$$

Thus, Equation 3.56 can be modified as:

$$N_{layer} = \frac{P_{sv}L_{avg}I_{pr}N_{li}N_d}{A_r R_e n_{tsv}} \text{ or } \max(\frac{S_{fi}R_i}{P_{di}}) \qquad (3.58)$$

where i represents a single device. Assume that the access connectivity of the chips in the MCM substrate is negligible compared to the interconnect connectivity. Substituting for the average routing length per interconnection using the uniform pin distribution technique and the interconnection-to-pin ratio in terms of the net size, the number of signal layers can be approximated to:

$$N_{layer} \approx \frac{0.34 P_{sv} N_d N_{li}}{n A_r^{0.5} R_e n_{tsv}}$$

(3.59)

Example: To illustrate, consider a printed wiring board with different types of device configurations. The input substrate and interconnection information are presented in Table 3.8, and the input device information is presented in Table 3.9.

Table 3.8. Input substrate information: example application

Parameter	Value
Substrate length	7.5-in.
Substrate width	8.5-in.
Pitch between signal vias	0.100-in.
Signal trace width	0.012-in.
Signal trace spacing	0.012-in.
Signal via pad diameter	0.060-in.
Average net size	2
Number of signal interconnections	1146

Table 3.9. Input device information: example application

Package style	Lead pitch (in.)	Length (in.)	Width (in.)	Number of of leads	Number of devices
AXL-res/cap	0.100	0.400	0.100	2	5
DIP	0.100	0.700	0.300	14	1
DIP	0.100	1.600	0.600	32	1
DIP	0.100	3.200	0.800	64	1
PGA	0.100	1.000	1.000	81	1
PGA	0.100	1.800	1.800	256	1
PGA	0.100	2.080	2.080	400	1
PLCC	0.050	0.550	0.550	44	1
PLCC	0.050	1.050	1.050	68	1
PLCC	0.050	1.650	1.650	84	1
QFP	0.020	1.000	1.000	200	1
QFP	0.020	1.500	1.500	300	1
QFP	0.020	2.100	2.100	400	1
ECONN	0.100	3.200	0.400	32	1
HCONN	0.100	3.200	0.200	64	1
HCONN	0.100	3.200	0.300	96	1

The total substrate area is 63.75 sq.in. and the routable area is 60 sq.in. The total device area is 27.89 sq.in. and the area utilization factor is determined from Equation 3.25 as 0.46. The average routing length per interconnection determined by the uniform pin distribution technique is 2.641 in. Consider, for example, the PGA package with 256 leads. The permitted connectivity of the inner and outer layers of the PGA are the same because the PGA is a through-hole device. The permitted number of traces per channel with a trace width of 0.012 in., trace spacing of 0.012 in., via-pad diameter of 0.060-in. and via-pitch of 0.100 in. is calculated as one. Thus, the permitted connectivity ($C_{p-PGA256}$) is found to be 10 in./sq.in. From the device information, the number of rows of pins for the PGA is found to be five. The shape factor of the PGA is obtained from Table 3.7 as two and the access connectivity ($C_{a-PGA256}$) is 100 in./sq.in.

◇

Table 3.10. Connectivity values: example application

Package style (i)	$C_{pi-outer}$	$C_{pi-inner}$	C_{ai}			C_{ti}	
			S_{fi}	R_i	C_{ai}	N_{li}	C_{ti}
AXL	10	10	1	1	10	2	30.62
DIP	10	10	1	1	10	14	40.83
DIP	10	10	1	1	10	32	20.41
DIP	10	10	1	1	10	64	15.31
PGA	10	10	2	3	60	81	49.60
PGA	10	10	2	5	100	256	48.39
PGA	10	10	2	8	160	400	56.62
PLCC	10	10	2	1	40	44	89.07
PLCC	10	10	2	1	40	68	57.64
PLCC	10	10	2	1	40	132	29.69
QFP	0	10	2	1	100	200	122.4
QFP	0	10	2	1	100	300	81.65
QFP	0	10	2	1	100	400	58.32
ECONN	10	10	1	1	10	32	15.31
HCONN	10	10	1	2	20	64	61.24
HCONN	10	10	1	3	30	96	61.24

The interconnection-to-pin ratio of the substrate from Equation 3.29 is 0.5. With an average routing length of 2.641 in., a device area of 3.24 sq.in., an area utilization factor of 0.46 and an interconnection-to-pin ratio of 0.50, the interconnect connectivity ($C_{t-PGA256}$) is calculated as 48.39 in./sq.in., using Equation 3.48. Similarly, the connectivity values of the other devices are computed and the results shown in Table 3.10. Among the various devices, PGAs and QFPs with high lead-counts and low lead pitches (PGA-256, QFP-200, QFP-300, PGA-400) demand the highest connectivity.

The average permitted connectivity of the outer layer is 9.05 in./sq.in. and the average permitted connectivity of the inner layers is 10 in./sq.in. The average interconnect connectivity is 57.22 in./sq.in. and the maximum access connectivity is 160 in./sq.in. The average number of signal layers for a routing efficiency of 50 percent is calculated as:

$$
\begin{aligned}
N_{layer} &= \frac{1}{R_e}\text{ave}(\frac{C_{di} - C_{pi-outer}}{C_{pi-inner}} + 1) \text{ or } \max(\frac{C_{ai}}{C_{pi-outer}}) \\
&= \frac{1}{0.5}(\frac{41.9 - 9.05}{10} + 1) \approx 10 \text{ or } (\frac{160}{9.05}) \approx 16 \qquad (3.60)
\end{aligned}
$$

Thus, the substrate is estimated to have 16 signal layers. Another advantage of the estimation model is the facility to study the effect of factors such as the routable substrate area, the number of signal interconnections and the conductor dimensions on the number of signal layers. Figure 3.15 shows the variation of the number of layers with the interconnection-to-pin ratio and the area utilization factor. Figure 3.16 shows the variation of the number of layers with the number of traces per channel. Assume that the number of signal interconnections remains constant for the substrate and that the designer is mainly concerned with the trade-offs between device configuration, the number of traces per channel and the routable substrate area. The lowest number of signal layers that can be obtained with an area utilization factor of 10 percent (routable substrate area: 278.9 sq.in.) is eight and with four traces per channel is two. Thus, the effect of the number of traces per channel is more significant because of the presence of the PGA and QFP packages (PGA-256, PGA-400, QFP-300, QFP-400), where the access connectivity exceeds the interconnect connectivity. Since access connectivity does not depend upon the routable substrate area and is only dependent on the device configuration and the conductor dimensions, increases in the routable area do not reduce the required number of signal layers. The replacement of PGA-256, PGA-400, QFP-300 and QFP-400 will result in lower demanded connectivity, but substitution with devices of low pin density will result in increased routable substrate area. Figure 3.16 shows that the required number of signal layers is reduced from 12 to 4 when the number of traces per channel is increased from one to two. Thus, in the example application, the number of traces per channel is the primary factor for reducing the number of signal layers.

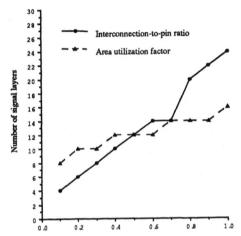

Figure 3.15. Signal layers vs. interconnection-to-pin ratio
and area utilization factor: example application

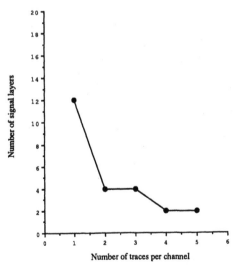

Figure 3.16. Signal layers vs. number of traces per channel: example application

3.4 References

[CHE80]   Cheng, S., "Statistics of Wire Lengths on Circuit Boards", *IEEE Transactions on Components, Hybrids and Manufacturing Technology*, vol. 3, no. 2, June 1980, pp. 305-306.

[COO88]   Coombs, C. F., *Printed Circuits Handbook*, McGraw Hill, New York, 3rd edition, 1988.

[DON79]   Donath, W. E., "Placement and Average Interconnection Lengths of Computer Logic," *IEEE Transaction on Circuits and Systems*, vol. CAS-26, April 1979, pp. 272-277.

[HAR89]   Hardenberg, W., "PWB Design," *Electronics Materials Handbook*, 1989, pp. 110-113.

[MAR88]   Marsh, M. G., *Pads-Superrouter*, Cad Software Inc., June 1988.

[MAR89]   Marsh, M., "Why doesn't my Autorouter Work?", *Circuit Design*, October 1989, pp. 8-14.

[MES87]   Messner, G., "Cost-density Analysis of Interconnections," *IEEE Transactions on Components, Hybrids and Manufacturing Technology*, vol. CHMT-10, no. 2, June 1987, pp. 143-151.

[MUR90]   Murphy, D. W., and Murvihill, W. D., "Characteristics of Four Large PCBs," *Circuit Design*, May 1990, pp. 5-13.

[NAK90]   Nakamura, K., "Packaging Technologies of Large-Scale General Purpose Computers," *Electronic Engineering Times*, December 10, 1990.

[SER78]   Seraphim, D. P., "Chip-Module Package Interfaces," *IEEE Transactions on Components, Hybrids and Manufacturing Technology*, vol. CHMT-1, no. 3, September 1978, pp. 305-309.

[TOL91]   Tollis, I. G., "A New Approach to Wiring Length," *IEEE Transactions on Computer Aided Design*, November 1991, pp. 1392-1400.

[TUM89]   Tummala, R., *Microelectronics Packaging Handbook*, Van Nostrand, Reinhold, 1989.

[BOG90]   Bogatin, E., "High Technology Performance Requirements," *Integrated Circuit Engineerin Corporation*, 1990, pp. 4-5 to 4-52.

Chapter 4

Placement for Routability

Yeun Tsun Wong, Michael Pecht, Michael D. Osterman and Guoqing Li

Placement involves positioning modules on a workspace with respect to some measures of success while observing specifications and constraints. Good placement configurations meet performance requirements while providing easy and neat producibility, high reliability, and low cost.

Historically, placement techniques have been developed on the basis of minimizing the total wire length on the workspace as the fundamental factor for routability and perceived performance [STE61, BRE72, HAN72, HU85, OHT86, PRE86, PEC91]. With the advent of high speed technologies, restricting the maximum length of signal propagation critical routing paths, clustering functionally related modules to conform to speed and transmission line requirements, designing signal paths to meet required electrical characteristics (resistance, current capacity, and capacitances), restricting tolerances on congestion, and keeping analog and digital functions shielded to prevent cross talk, also became fundamental routing and performance considerations.

Because it is impossible to incorporate all of the desired attributes of placement for routability into a coherent mathematical model, a cost function expressing the main routability features is usually employed, often using heuristic methods. Minimizing the cost computed with the cost function is considered the placement objective. There are two major goals in placement for routability:

- formulating the placement objectives consistent to the routing objective;
- determining the placement process for minimizing the cost without falling in local minima.

The placement objective should be independent of the initial placement and the layout complexity, otherwise, the placement process is divergent, and thus it is difficult to meet the placement objective [HAN76, HAR86, WON89].

In this chapter, cost functions and placement processes will be introduced. Constructive placement methods are presented for background informations. Iterative placement methods are then determined with state-of-art methods. Applying the theorem proven in Chapter 2 on finding a minimum spanning tree from the graph partitioning, measures to compare interconnection configuration are examined. In particular, the effect of tree type choice on the placement objective measured by the minimum total length of minimum rectilinear Steiner trees (MRSTs) is discussed, and an approach to formulate a placement objective which coincides with routing objective is presented.

4.1 Cost Functions

A routability cost function is a mathematical means of measuring the routability of a given stage in the placement process. Generally, the cost function, C, is defined by two terms: a weighted routing (wire) length, W, or equivalent relation pertaining to signal propagation, and a correction term, W_c, employed for improving the placement configuration:

$$C = W + W_c \qquad (4.1)$$

To compute the cost function, the size and shape of the modules, the terminal slots in a module, the interconnection tree corresponding to the signal set, and the special requirement for the signal set must be considered. In an iterative placement, the cost reduction, computed from the cost function for two placement configurations, is often employed to determine which configuration should be accepted, if the redundant computations of costs for different placement configurations can be avoided. Maximizing the cost reduction can also be considered to be the placement objective.

4.1.1 Routing length computations

Routing length depends on the type of interconnection tree employed (see Chapter 2). If each net is connected as a specific tree type, the common model for the total weighted tree length can be computed by Equation 4.2:

$$W = \sum_{i=1}^{n_m} w_i l_i \qquad (4.2)$$

where w_i is the weighting factor of the ith of n_m nets of length l_i. In the same net, the weighting factor for horizontal wires may be distinct from that of the vertical wires. Therefore, $w_i l_i$ can be decomposed to:

$$w_i l_i = w_{hi} l_{hi} + w_{vi} l_{vi} \qquad (4.3)$$

where w_{hi} and w_{vi} are weighting factors for the horizontal wires and vertical wires and l_{hi} and l_{vi} are the total length of horizontal wires and the total length of vertical wires of the ith net, respectively. Based on performance and manufacturing requirements, a priority order may be given to a set of nets, to horizontal routing segments, or to vertical routing segments. A net with a higher priority should have a larger weighting factor, so that the shorter routing length for this net can be obtained from the placement.

The interconnection tree is made up of vertices (terminals) that are connected by edges (routing wires). Usually, the minimum spanning tree length is employed to measure the tree length. To speed up the placement process, the approximate length of a minimum spanning tree can be obtained from the length of an equivalent graph, such as a complete graph in which every terminal is connected to other terminals in the same signal set, or from the minimum rectangle enclosing all terminals of a signal net with the minimum perimeter length.

For a signal set $\{v_1, ..., v_n\}$ in which v_i $(1 \leq i \leq n)$ is a terminal, the average edge length in the complete graph connecting $\{v_1, ..., v_n\}$ is $\sum_{i=1}^{n-1} \sum_{j=i+1}^{n} l_{v_i v_j}/C(n,2)$, where $l_{v_i v_j}$ is the length of edge $e_{v_i v_j}$. For a tree with $n-1$ edges, the average tree length, l_c, computed from the complete graph is:

$$
\begin{aligned}
l_c &= \frac{n-1}{C(n,2)} \sum_{i=1}^{n-1} \sum_{j=i+1}^{n} l_{v_i v_j} \\
&= \frac{2}{n} \sum_{i=1}^{n-1} \sum_{j=i+1}^{n} l_{v_i v_j} \qquad (4.4)
\end{aligned}
$$

The time for computing l_c is proportional to $n(n-1)$.

If the minimum rectangle, R, that encloses $\{v_1, ..., v_n\}$ is expressed by the lower-left and upper-right vertices, $v_{ll} = (x_{v_{ll}}, y_{v_{ll}})$ and $v_{ur} = (x_{v_{ur}}, y_{v_{ur}})$, then $x_{v_{ll}} = min(x_1, ..., x_n)$, $y_{v_{ll}} = min(y_1, ..., y_n)$, $x_{v_{ur}} = max(x_1, ..., x_n)$, and $y_{v_{ur}} = max(y_1, ..., y_n)$, where max (or min) is a selection function that chooses a maximum (or minimum) value from the following parameter list. When the tree length, l_h, is approximated from the half perimeter of R, then:

$$l_h = [(x_{ur} - x_{ll}) - (y_{ur} - y_{ll})] \tag{4.5}$$

The computation for the half-perimeter method is more efficient than that for the complete graph method, because the time for computing l_h is linear to n.

From the experimental data reported by A. Goto and T. Matsuda [GOT86], when the number of terminals in every net is small (such as those in LSI circuits), the correlations between the minimum rectilinear spanning tree length, l_{MSPT}, and l_c (rectilinear) and between l_{MSPT} and l_h can be estimated by

$$l_c = (0.9l_{MSPT} + 715) \pm 24 \tag{4.6}$$

and

$$l_h = (0.64l_{MSPT} + 1100) \pm 9 \tag{4.7}$$

where $1400 \leq l_{MSPT} \leq 2100$. Because $max(\Delta l_c) = 48$ and $max(\Delta l_h) = 18$, using the half-perimeter method in the cost function yields a better approximation.

4.1.2   Correction functions

A correction function, W_c, sometimes called a penalty function, can be used to indicate the local routing congestion, δ_1, measured by the number of wires per workspace area, module overlap, δ_2, and other routing constraints:

$$W_c = \sum_i \delta_i \tag{4.8}$$

Let A_i be the area of a rectangle enclosing net i with the minimum perimeter and P_a be the possibility of occupying each slot $a \in A_i$ by net i. The larger overlap area of A_1 to A_n indicates the higher wire density for placement with n nets, and the large $(\sum_{i=1}^n A_i - \bigcup_{i=1}^n A_i)$ means the local congestion is more severe.

Considering $P_a = P_b$ for arbitrary $a, b \in A_i$ $(i = 1, ..., n)$, the correction function for routing congestion can be computed by:

$$\delta_1 = \alpha_1(\sum_{i=1}^{n} A_i - \bigcup_{i=1}^{n} A_i) \qquad (4.9)$$

where α_1 is a penalty length per unit area. The large δ_1 will result in a longer routing length that causes a rejection of configuration with the routing congestion, or provides more area (longer routing length) for digesting the routing congestion provided that the configuration with module overlap is accepted.

Considering the total module overlap area to be A_o, the correction function for module overlap can be computed by:

$$\delta_2 = \alpha_2 A_o \qquad (4.10)$$

where α_2 is a penalty length per unit overlap area. The larger δ_2 will result in a longer routing length that will either reject the configuration with module overlaps or provide more area (longer routing length) for separating overlap modules in the event that the configuration with module overlap is accepted. However, if two modules are separated by an area greater than required, then $A_o < 0$. Modules with non-uniformed sizes and shapes can be placed using δ_2.

4.1.3 Partitioning pertaining to placement objectives

To compute a cost function hierarchically, modules are often partitioned into subsets by a partition boundary [FIS67, KER70, COR79, FID82]. Partioning in placement can be divided into two kinds: depth-first and breadth-first. In depth-first partitioning, partitioning starts with every module, then every subset of two adjacent modules, every subset of three adjacent modules, etc., until finally two subsets of adjacent modules are partitioned. In contrast, breadth-first partitioning begins by partitioning all modules into two subsets, and then every subset is partitioned into two, until the subset contains one module. Breadth-first partitioning is most widely used in the iterative placement process.

With restrictions on the size (or area) of each module and the number of external nets that crosses two different sets of modules, mathematically, the partition problem is defined as follows: Let V be a set of modules, $S(v)$ be the size of $v \in V$, $N(v)$ be the number of nets connected to v, and I be the number of nets that connect $v \in V_i \subset V$ with all other modules in V_i. The number of external nets for v is defined as:

$$E(v) = N(v) - I(v) \tag{4.11}$$

By defining the size limit and the external net limit as S_{max} and E_{max}, respectively, the partition problem becomes a process of finding subsets of V pertaining to a module configuration, such that:

$$\sum_{v \in V_i} S(v) \leq S_{max} \tag{4.12}$$

and

$$\sum_{v \in V_i} E(v) \leq E_{max} \tag{4.13}$$

where $V_i \cap V_j = \emptyset$ $(i \neq j)$ and $V_i \subset V$. If the objective is to minimize the total number of external nets among partitioned sets $V_1, ..., V_n$, then

$$E_{min} = min(\frac{1}{2} \sum_{V_i \subset V} \sum_{v \in V_i} E(v)) \leq E_{max} \tag{4.14}$$

where E_{min} is the desired total number of external nets. In the "min-cut" placement method (see Section 4.3.3), $I(v)$ and $E(v)$ are employed to obtain E_{min}.

Similarly, if we define the shortest edge connecting $v \in V_i$ and $u \in (V - V_i)$ as an external edge and the total length of the external edge connecting v as $L(v) = E(v)l(v)$, where $l(v)$ is the average length of edges connecting v, then the partition problem can also become a process of finding subsets of V pertaining to a module configuration, such that $\sum_{v \in V_i} L(v)$ is the minimum. According to Corollary 2.2.1, the minimum total length of all minimum spanning trees, L_{MSPT}, can be obtained by minimizing the total length of all external edges. Therefore, finding L_{MSPT} can be converted to the partition problem:

$$\begin{aligned} L_{MSPT} &= min(\frac{1}{2} \sum_{V_i \subset V} \sum_{v \in V_i} L(v)) \\ &= min(\frac{1}{2} \sum_{V_i \subset V} \sum_{v \in V_i} E(v)l(v)) \\ &= min(\bar{l})E_{min} \end{aligned} \tag{4.15}$$

where $\bar{l}$ is the average length of all external edges. Equation 4.15 shows that L_{MSPT} can be obtained by minimizing both the total number of external nets and the average length of all external edges.

4.2   Constructive Techniques

Placement processes are often categorized as continuous and noncontinuous. Continuous techniques treat the workspace as a continuous surface on which the modules are free to reside, while noncontinuous techniques partition the workspace into slots to which the modules are assigned. The noncontinuous approach is extremely useful for workspaces in which most of the modules to be placed have similar shapes and sizes, because the geometric problems encountered in placement are simplified.

Placement algorithms employed in the solution of the placement problem, whether continuous or noncontinuous, fall into two classes: constructive or iterative. Constructive procedures are typically applied to establish an initial placement configuration (assignment). Iterative techniques, which seek to improve the placement configuration, operate on an initial placement configuration generated either randomly or by constructive procedures.

In constructive techniques, the placement configuration is usually determined by positioning unplaced modules onto a workspace containing previously placed modules [HAN72, SCH72]. Once modules are placed, they are not moved. The particular rules for selection and placement of the modules are determined by the specific constructive strategy. Generally, the selection of unplaced modules for placement and the determination of where to place the module(s) is a sequential process based on how "strongly" the unplaced module(s) is "bound" to the placed modules.

Constructive techniques involve partitioning modules into a set, A, of placed modules and a set, B, of unplaced modules. Their objective functions are based on the connectivity between two different modules $u \in A$ and $v \in B$, which is the number of nets connecting u and v, and a weighting factor which could express the need for certain modules to be closer together. Denoting the weighting factor as w_{uv}, the number of nets connecting u and v as c'_{uv} and the cost as $c_{uv} = w_{uv}c'_{uv}$, $v \in B$ is selected if v has the maximum weighted connectivity with $u \in A$:

$$f_1(v) = max(c_{uv}; \quad u \in A, \ v \in B) \qquad (4.16)$$

Another selection rule is based on selecting $v \in B$, which has the total maximum weighted connectivity with all $u \in A$:

$$f_2(v) = max(\sum_{u \in A} c_{uv}; \quad v \in B) \qquad (4.17)$$

where $\sum_{u \in A} c_{uv}$ is the number of external nets $E(v)$ for $v \in B$, given in Equation 4.11 when $w_{uv} = 1$. To avoid repetitively computing the cost in every placement stage by the cost function shown in Equation 4.16 or 4.17, a cost matrix can be employed, in which each element indicates a cost for two modules. For a workspace with n modules, the computational complexity for the cost function is $O(n^2)$.

Example: If signal sets $s_1 = \{a, b, e\}$, $s_2 = \{a, b, c, f\}$, $s_3 = \{c, d, f\}$, $s_4 = \{b, c, g\}$, $w_{uv} = 2$ when $\{u, v\} \subset s_1$, and $w_{uv} = 1$ when $\{u, v\} \subset \cup_{i=2}^4 s_i$, then the cost matrix of c_{uv} is shown in Table 4.1. Let $A = \{a, b, c\}$ and $B = \{d, e, f, g\}$. Employing $f_1(v)$, $v = e$ or $v = f$ is selected, because $C_{ea} = C_{eb} = C_{fc} = 2$ is the maximum weighted connectivity between $u \in A$ and $v \in B$. Employing $f_2(v)$, $v = e$ is selected, because $C_{ea} + C_{eb} = 4$ is the maximum weighted connectivity between $u \in A$ and $v \in B$.

Table 4.1.   Cost Matrix of c_{uv} $(u, v \subset \{a, b, c, d, e, f, g\})$

	a	b	c	d	e	f	g
a	0	3	3	0	2	1	0
b	3	0	2	0	2	1	1
c	3	2	0	1	0	2	1
d	0	0	1	0	0	1	0
e	2	2	0	0	0	0	0
f	1	1	2	1	0	0	0
g	0	1	1	0	0	0	0

◇

If the weighting factor is set to unity, then the selection criterion involves selecting the unplaced module connecting the most edges to a placed module, using Equation 4.16, or to all placed modules, using Equation 4.17. Generally, the method involves using Equation 4.16 to select a "seed" or initial module or a set of modules, then using Equation 4.17 to place the remaining modules.

Constructive methods are generally used only to prepare an initial placement for an iterative method. Three classes of constructive placement methods, pair-linking, cluster-development and quadratic assignment, will now be introduced. While each of these methods possess desirable characteristics, each also exhibits some undersirable features.

4.2.1 Pair-linking

The selection rule of the pair-linking method is based upon the cost given by Equation 4.16 or 4.17. Initially, a module, $u \in A$, is selected as a "seed" by Equation 4.16 and placed in the center of the workspace. The next selected module is placed in an unoccupied slot nearest to slot (x_v, y_v) and is determined by the slot condition:

$$f_3(u,v) = min(\sum_{u \in A} c_{uv}(|x_u - x_v| + |y_u - y_v|); \ v \in B) \tag{4.18}$$

From Equation 4.18, the slot for $v = (x_v, y_v)$ is given as the following:

$$x_v = \sum_{u \in A} c_{uv} x_u / \sum_{u \in A} c_{uv} \tag{4.19}$$

and

$$y_v = \sum_{u \in A} c_{uv} y_u / \sum_{u \in A} c_{uv} \tag{4.20}$$

Example: This example shows the "seed" selection and module placement for the cost matrix given in Table 4.1, using Equations 4.16 to 4.20. The cost matrix in Table 4.1 shows that the maximum cost is given by the edge between a and b or between a and c. Using $f_1(v)$, a is selected as a "seed" and placed in the center of the workspace. Equations 4.19 and 4.20 indicate that $x_b = c_{ab} x_a / c_{ab} = x_a$ and $y_b = c_{ab} y_a / c_{ab} = y_a$. Because b cannot be placed in the same slot with a, b is placed in a slot next to a. Similarly, c is selected and placed in a slot next to a. In this stage, $A = \{a, b, c\}$ and $B = \{d, e, f, g\}$. Because the maximum cost between $u \in A$ and $v \in B$ is given by the edge between e and a or between e and b, e is selected and placed in the slot nearest to a or b. In this stage, $A = \{a, b, c, e\}$ and $B = \{d, f, g\}$. Because the cost given by the edge between f and b is the maximum, f is placed in the slot nearest to b. Finally, d is placed in the slot nearest to c or f, and g is placed in the slot nearest to b or c.
◇

4.2.2 Cluster-development

The cluster-development method is similar to the pair-linking method, but the selection rule is different from those given in Equations 4.16 and 4.17. For every

partially placed signal set k, there exists at least one edge leading from the unplaced modules to the placed modules. If T_k is the maximum possible number of
edges leading from a module, and n_k is the number of modules in the kth signal
set, then the average number of edges is $\sum_{j=1}^{T_k} j/T_k = (1+T_k)T_k/(2T_k) = (1+T_k)/2$
where $T_k = n_k - 1$ for the signal net connected as the minimum spanning tree.
Considering the average number of modules in B that are connected to a module in A is $T_k/2$, if the number of signal sets common to v is n_v, then the total
expected number of edges per module in B is:

$$F(v) = \sum_{k=1}^{n_v} \frac{1+T_k}{T_k} \tag{4.21}$$

From the maximum of $F(v)$, $v \in B$ can be selected and placed in the slot (x_v, y_v),
which is determined by Equations 4.19 and 4.20.

4.2.3   Quadratic assignment

The quadratic assignment problem is formulated as follows. Given a cost matrix
$C = \{c_{ij}\}$, and a distance matrix $D = \{d(x_i, x_j)\}$, where c_{ij} is the connectivity
of modules i and j, and $d(x_i, x_j)$ is the distance between modules x_i and x_j, find
the minimum of total routing length

$$L = \frac{1}{2} \sum_{i,j=1}^{n} c_{ij} d(x_i, x_j) \tag{4.22}$$

over all permutations of assigning modules in $\vec{x} = [x_1, ..., x_n]$ [HAN72, CHE84,
BLA85, SUA88, KLE88, SIG91, TSA91]. To confine the modules in slots, m_1,
m_2, ..., m_n, the following slot constraints must be satisfied:

$$\sum_{i=1}^{n} x_i \;\; = \;\; \sum_{i=1}^{n} m_i = a_1 \tag{4.23}$$

$$\sum_{i=1}^{n} x_i^2 \;\; = \;\; \sum_{i=1}^{n} m_i^2 = a_2 \tag{4.24}$$

$$\cdots$$

$$\sum_{i=1}^{n} x_i^n \;\; = \;\; \sum_{i=1}^{n} m_i^n = a_n \tag{4.25}$$

where Equations 4.23 and 4.24 are called the first-order and the second-order
constraints, respectively. In this section, a methodology developed by R-S. Tsay
and E. Kuh [TSA91] is introduced.

Considering nets to be two-terminal nets and modifying D to $B = \{b_{ij}\}$, where

$$b_{ij} = \begin{cases} -c_{ij} & \text{if } i \neq j, \\ \sum_{k=1}^{n} c_{ij} & \text{if } i = j, \end{cases} \qquad (4.26)$$

then the vector of the total routing length with the Euclidean distance can be expressed as:

$$L(\vec{x}) = \frac{1}{2} \sum_{i,j=1}^{n} c_{ij}(x_i - x_j)^2 = \vec{x}^T B \vec{x} \qquad (4.27)$$

Because B is real and symmetric, it can be expanded into the outer product of the eigenvector:

$$B = \sum_{i=0}^{n-1} \lambda_i \vec{v}_i \vec{v}_i^T \qquad (4.28)$$

where $\vec{v}_i$ is a unit vector and is the ith eigenvector with respect to λ_i. Using eigenvectors of B, $\vec{x}$ can be expressed:

$$\vec{x} = \sum_{i=0}^{n-1} \alpha_i \vec{v}_i = \alpha_0 \vec{v}_0 + \tilde{x} \qquad (4.29)$$

where $\lambda_0 = 0$, $\vec{v}_0 = [1, ..., 1]^T / \sqrt{n}$, $\alpha_0 = \vec{x}^T \vec{v}_0 = a_1 / \sqrt{n}$ and $\tilde{x} = \sum_{i=1}^{n-1} \alpha_i \vec{v}_i$. Because $\lambda_0 = 0$, an equivalent problem can be defined on the space orthogonal to $\vec{v}_0$ with the basis $\{v_1, v_2, ..., v_{n-1}\}$:

$$L(\vec{x}) = L(\tilde{x}) = \sum_{i=1}^{n-1} \lambda_i \alpha_i^2 \qquad (4.30)$$

According to Equation 4.29 and $a_2 = \sum_{i=1}^{n-1} \alpha_i^2$,

$$|\tilde{x}|^2 = \sum_{i=1}^{n-1} \alpha_i^2 = |\vec{x}|^2 - |\alpha_0 \vec{v}_0|^2 = a_2 - \alpha_0^2 = a_2 - a_1^2/n \qquad (4.31)$$

According to Equation 4.31,

$$\alpha_1^2 = -\sum_{i=2}^{n-1} \alpha_i^2 + (a_2 - a_1^2/n) \qquad (4.32)$$

Multiplying Equation 4.32 by λ_1 and adding it to Equation 4.30 yields:

$$L - \lambda_1(a_2 - a_1^2/n) = \sum_{i=2}^{n-1} (\lambda_i - \lambda_1)\alpha_i^2 \qquad (4.33)$$

or

$$\sum_{i=2}^{n-1} \frac{\alpha_i^2}{r_i^2} = 1 \qquad (4.34)$$

where $r_i = \sqrt{[L - \lambda_1(a_2 - a_1^2/n)]/(\lambda_i - \lambda_1)}$. An arbitrary point $(\alpha_2, \alpha_3, ..., \alpha_{n-1})$ that satisfies Equation 4.34 is on the surface of an $(n-2)$-dimensional ellipsoid with $\vec{x}_1 = \sqrt{a_2 - a_1^2/n}\ \vec{v}_1$ as the center and with $r_2, ..., r_{n-1}$ as the radius for a given $L(\vec{x})$. The center point is the solution of the problem, if it satisfies the constraint:

$$\sum_{i=2}^{n-1} \alpha_i^2 \leq a_2 - \frac{a_1^2}{n}. \qquad (4.35)$$

Because this solution only satisfies the first-order and second-order slot constraints, and only x_1 is determined, the solution has to further satisfy all other high-order slot constraints by the subsequent process for selecting the module slots nearest to the center point of the ellipsoid.

4.3   Iterative Techniques

In iterative placement methods, modules are selected and moved to alternate slots based on the new slot's cost (reduction) computed from a cost (reduction) function. If the resulting configuration is better than the old, the new configuration is retained; otherwise the previous configuration is restored. This process continues until some stopping criterion is met. The stopping criterion is based on a relative or absolute improvement in the evaluation metric, or some fixed number of iterations. The generic form for an iterative placement algorithm is:

> *PROCEDURE PLACEMENT (net list, modules and workspace physical data);*
> *BEGIN*
> > *Compute* **COST** *for initial placement;*
> > *UNTIL stopping criterion is satisfied DO*
> > > **SELECT** *module(s) to move;*
> > > **MOVE** *selected module(s) to trial slots;*
> > > *Compute* **COST** *for trial placement;*
> > > *IF trial* **COST** < *current* **COST** *THEN*
> > > > *current* **COST** = *trial* **COST**
> > > *ELSE* **MOVE** *selected module(s) to previous slots*
> > *ENDLOOP;*
> *END.*

In this algorithm, the **COST** is computed from a cost function and **SELECT** and **MOVE** determine the iterative process. Once the modules are selected, the movement function, which often swaps two selected modules, determines new slots for those modules. After the selected modules are moved to a new slot, a cost computed from the cost function measures the quality of the new arrangement. In order to control iterative processes without falling into local minima, following basic techniques for selection and movement, which tend to keep the interconnection system at the low energy state or obtain the maximum total cost reduction accumulated from all iterative steps, are often used:

- force techniques simulating the minimum tension state of the attractive force system;
- simulated annealing techniques simulating the thermal equilibrium state;
- maximum accumulated-cost-reduction techniques.

4.3.1 Force-directed placement

Force-directed placement procedures are based on the principle that fictitious forces proportional to the distance between modules can be used to minimize the routing length [HAN72, QUI79, FOR87]. The goal of this technique is to find the relative position of each module on the workspace that will result in a zero force. The general force-directed equations are modeled after Hooke's law, which states that if two particles are connected to each other by a spring, the attractive force between the two particles equals to the product of the spring constant and the distance between them. For placement, the modules are considered to be particles, and the spring constant is a function of the number of signal sets common to the two modules. The goal is then to minimize the total routing length. The force-directed placement proposed by N. R. Quinn, Jr., and M. A. Breuer is introduced as follows [QUI79]:

Let (x_i, y_i) be the coordinates of modules i, $\Delta S_{ij} = -\Delta S_{ji}$ be the Manhattan distance of modules i and j, and k_{ij} be the number of signal nets common in modules i and j. Then

$$|\Delta S_{ij}| = |\Delta x_{ij}| + |\Delta y_{ij}| \qquad (4.36)$$

and the force, f_{ij}, acting on module i by module j is expressed as:

$$f_{ij} = -k_{ij}\Delta S_{ij} \qquad (4.37)$$

where $f_{x_{ij}} = -k_{ij}\Delta x_{ij} = -k_{ij}(x_i - x_j)$, and $f_{y_{ij}} = -k_{ij}\Delta y_{ij} = -k_{ij}(y_i - y_j)$.

Because x_i and y_i are constants in the solution of $f_{ij} = 0$, all modules collapse to a single point. To separate modules, a force of repulsion, R_{ij}, is required, which is typically a function of number of movable modules, M, as well as the interconnection topology. When $k_{ij} = 0$ or $i = j$, $R_{ij} = 0$. When $k_{ij} \neq 0$, it is suggested that:

$$R_{ij} = \frac{\sum_{i=1}^{M}\sum_{j=1}^{M} k_{ij}}{C_R n_T} \tag{4.38}$$

where $1 \le C_R \le 2$ and n_T is the number of $k_{ij} = 0$ terms. With the balance of the repulsion, the force equation can be modified to:

$$P_{ij} = -k_{ij}\Delta S_{ij} + R_{ij} \tag{4.39}$$

Along the x and y directions, Equation 4.39 can be decomposed into

$$P_{x_{ij}} = P_{ij}\frac{\Delta x_{ij}}{|\Delta S_{ij}|} = -k_{ij}\Delta x_{ij} + R_{ij}\frac{\Delta x_{ij}}{|\Delta S_{ij}|} \tag{4.40}$$

and

$$P_{y_{ij}} = P_{ij}\frac{\Delta y_{ij}}{|\Delta S_{ij}|} = -k_{ij}\Delta x_{ij} + R_{ij}\frac{\Delta y_{ij}}{|\Delta S_{ij}|} \tag{4.41}$$

If M of total N modules is movable, Equations 4.40 and 4.41 become:

$$P_{x_i} = \sum_{j=1}^{N} P_{x_{ij}} \tag{4.42}$$

and

$$P_{y_i} = \sum_{j=1}^{N} P_{y_{ij}} \tag{4.43}$$

where $i = 1, ..., M$.

To avoid an unacceptable result in which movable modules tend to surround the connectors, it is required to maintain the center of a set of movable modules at a fixed slot. The force acting on the movable modules by the fixed modules are called the force on the center of mass, F_{CM}, with components F_{CM_x} and F_{CM_y}, which equals to the sum of forces acting on all movable modules, because the sum of forces acting on the center by all movable modules are balanced when the system of movable modules reaches the minimum tension state.

$$F_{CM_x} = \sum_{i=1}^{M} \sum_{j=M+1}^{N} -k_{ij}\Delta x_{ij} = \sum_{i=1}^{M} P_{x_i} \tag{4.44}$$

and

$$F_{CM_y} = \sum_{i=1}^{M} \sum_{j=M+1}^{N} -k_{ij}\Delta y_{ij} = \sum_{i=1}^{M} P_{y_i} \qquad (4.45)$$

Therefore, the solution can be obtained from the equations:

$$F_{x_i} = P_{x_i} - \frac{F_{CM_x}}{M} = \sum_{j=1}^{N}(-k_{ij}\Delta x_{ij} + R_{ij}\frac{\Delta x_{ij}}{|\Delta S_{ij}|}) - \frac{F_{CM_x}}{M} = 0 \qquad (4.46)$$

and

$$F_{y_i} = P_{y_i} - \frac{F_{CM_y}}{M} = \sum_{j=1}^{N}(-k_{ij}\Delta y_{ij} + R_{ij}\frac{\Delta y_{ij}}{|\Delta S_{ij}|}) - \frac{F_{CM_y}}{M} = 0 \qquad (4.47)$$

where $i = 1, ..., M$.

To iteratively solve the above equations, the following modifications are made:

$$\frac{\partial F x_i}{\partial x_i} = F'_{x_i} = \sum_{j=1}^{N} -k_{ij} + R_{ij}\frac{\Delta y_{ij}}{\Delta S_{ij}^2} + \frac{1}{M}\sum_{j=M+1}^{N} k_{ij} \qquad (4.48)$$

and

$$\frac{\partial F y_i}{\partial y_i} = F'_{y_i} = \sum_{j=1}^{N} -k_{ij} + R_{ij}\frac{\Delta x_{ij}}{\Delta S_{ij}^2} + \frac{1}{M}\sum_{j=M+1}^{N} k_{ij} \qquad (4.49)$$

where $\Delta x_{ij} = \Delta x_i = x_i - x_{i-1} = -F_{x_i}/(2F'_{x_i})$ and $\Delta y_{ij} = \Delta y_i = y_i - y_{i-1} = -F_{y_i}/(2F'_{y_i})$. Note that each module is moved $1/2$ the designated distance because two modules are attracted to each other by the same force. Based on Equations 4.42 to 4.49, given a set of initial values of x_i and y_i $(i = 1, ..., M)$, the solution can be iteratively solved by the procedure below:

REPEAT

    FOR $i = 1$ TO M DO BEGIN

        $P_{x_i} = \sum_{j=1}^{N} -k_{ij}\Delta x_{ij} + R_{ij}\Delta x_{ij}/|\Delta S_{ij}|$;

        $P_{y_i} = \sum_{j=1}^{N} -k_{ij}\Delta y_{ij} + R_{ij}\Delta y_{ij}/|\Delta S_{ij}|$;

        $F'_{x_i} = \sum_{j=1}^{N} -k_{ij} + R_{ij}\Delta y_{ij}/\Delta S_{ij}^2 + \sum_{j=M+1}^{N} k_{ij}/M$;

        $F'_{y_i} = \sum_{j=1}^{N} -k_{ij} + R_{ij}\Delta x_{ij}/\Delta S_{ij}^2 + \sum_{j=M+1}^{N} k_{ij}/M$;

    END;

    $F_{CM_x} = \sum_{i=1}^{M} P_{x_i}$;

    $F_{CM_y} = \sum_{i=1}^{M} P_{y_i}$;

$FOR\ i = 1\ TO\ M\ DO\ BEGIN$

$\qquad F_{x_i} = P_{x_i} - F_{CM_x}/M;$

$\qquad F_{y_i} = P_{y_i} - F_{CM_x}/M;$

$\qquad x_i = x_i - F_{x_i}/(2F'_{x_i});$

$\qquad y_i = y_i - F_{y_i}/(2F'_{y_i});$

$END;$

$UNTIL\ \epsilon < \sum_{i=1}^{M}[|F_{x_i}| + |F_{y_i}|];$

In this procedure, ϵ is a stopping criterion value.

Because a force acts on every pair of modules common in the same signal net, the routing length for each signal set equals to the length of the complete graph connecting all modules in this signal net. Therefore, the total routing length used in the force-directed technique is the total complete graph length.

4.3.2   Simulated annealing

Based on the initial cost, an "annealing schedule" specifies ending cost and the rate at which the cost is lowered [KIR83, LUN83, VEC83, ROM84, WHI84, NAH85, ROM85, NAH86, KLI87, GRO87, LAM88, SEC88, BUI89, JOH89, KLI90, ROS90, SAA90]. The simulated annealing process begins with a random initial placement. An altered placement is generated, and the resulting change in cost, ΔC, is computed by the cost function shown in Equation 4.1. If $\Delta C < 0$, then the move is accepted. If $\Delta C \geq 0$, then the move is accepted with probability $e^{-\Delta C/T}$, where T expresses the annealing schedule. Denoting the initial module configuration and the initial annealing schedule as S_0 and T_0 respectively, the simulated annealing placement methods proposed by S. Kirkpatrick, et al, [KIR83] can be expressed as the following procedure:

$PROCEDURE\ SimulatedAnnealing(S_0, T_0);$

$BEGIN$

$\qquad S = S_0;$

$\qquad T = T_0;$

$\qquad k = 0;$

$\qquad REPEAT$

$\qquad\qquad REPEAT$

SELECT and MOVE modules to obtain S' randomly;

$\Delta C = C(S') - C(S);$

IF $\Delta C < 0$ OR random number $r < e^{-\Delta C/T_k}$ $(0 \leq r \leq 1)$ THEN

$\quad C(S) = C(S')$

ELSE MOVE selected modules to previous positions;

UNTIL equilibrium is reached; {No module qualifies to be moved in T_k}

INC(k) and $T_k = \gamma_k T_{k-1}$; {$0 < \gamma_k < 1$ is an annealing coefficient}

UNTIL stopping criterion is satisfied;

Output solution S;

END.

In this procedure, cost $C(S)$ corresponds to module configuration S. In the annealing schedule, γ_k is typically $0.8 \sim 9.5$ and $\gamma_1 \approx 0.8 \sim 0.85$. T_0 is determined by a sequence of random moves and the average cost change per move, $\overline{\Delta C}$. A reasonable probability of acceptance at T_0 satisfies $e^{-\overline{\Delta C}/T_0} = P \approx 1$. This suggests that $T_0 = -\overline{\Delta C}/\ln(P)$.

At each T, moves are attempted until either there are mn downhill moves or the total number of moves exceeds $2mn$, where m is a user-specific constant, and n is the total number of modules. The annealing process is terminated when the number of accepted moves is less than, say 5%, of all moves made at a constant T, or when T is lower than a specific value. If the run time for the placement process is not limited, the resulting placement obtained by the annealing technique is better than other placement techniques. However, if the run time is limited to that consistent with the force-directed or min-cut technique, the resulting placement obtained by the annealing technique is no better than that obtained by the other techniques [NAH85].

4.3.3 Min-cut

Min-cut algorithms divide the modules into two sets, and attempt to minimize the number of weighted edges between the sets, under the constraint of each module area [KER70, BRE77, KHO77, LAU79, KOZ83, DUN85, BUI89, WEI89, VIJ89]. This process is repeated until the number of modules in each subset is much smaller than the total number of modules. The min-cut problem can, thus, be considered a partition problem like that given in Equation 4.14.

Given two module subsets, A and B, of a net N, if module $u \in A$ is the only module of N in A, then moving u to B reduces the cut number by 1. The reduced cut indicates that the edge connecting $u \in A$ and $v \in B$ has been removed. Similarly, if all modules of a net are in one set, say A, then moving one module of the net to B causes the net to be cut, and thus the cost is increased by 1.

Let $I(u)$ be the number of nets connecting u that have all their modules in A, and $E(u)$ be the number of nets connecting u when u is a module not connected to any other module in A. Moreover, let $D_x = E(x) - I(x)$ for all $x \in S$, where S is the workspace, and D_x is defined as the difference between external cost $E(x)$ and internal cost $I(x)$. The cost reduction, ΔC_{uv}, obtained by exchanging $u \in A$ and $v \in B$ is

$$\Delta C_{uv} = D_u + D_v - \delta_{uv} \qquad (4.50)$$

where δ_{uv} is a correction that avoids double counting when u and v are both in the same net.

Example: Consider the example of modules a and b in Figure 4.1. The result of the exchange is shown in Figure 4.2. Because a is external to net 2 and internal to net 1, $D_a = E(a) - I(a) = 1 - 1 = 0$. Because b is internal to net 3, $D_b = E(b) - I(b) = 0 - 1 = -1$. Because only net 2 connects to a and b, $\delta_{ab} = 1$. Thus, the cost reduction of exchanging a and b in this example is $\Delta C_{ab} = D_a + D_b - \delta_{ab} = 0 - 1 - 1 = -2$. The negative value means the cost is increased.

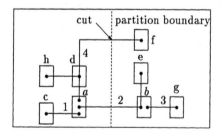

Figure 4.1. Partitioning and cuts
(Number of cuts=2)

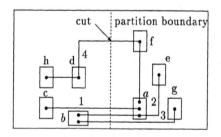

Figure 4.2. The cost reduction
obtained by exchanging a and b
(Cost reduction $= 2 - 4 = -2$)

⋄

In every pass of the iterative process, each selection of a pair $u \in A$ and $v \in B$ continues until all modules are selected. In each selection, if the cost is reduced by exchanging u and v, the exchange is accepted and the total cost is reduced. However, the fact that the total cost is increased by an exchange doesn't mean that the total cost cannot be reduced by subsequent exchanges. When the cost is reduced to a specific value and then is increased by a subsequent exchange, this specific value is called a local minimum. To avoid the placement process terminating in a local minimum, B. W. Kernighan and S. Lin proposed this min-cut procedure:

> *PROCEDURE MinCut(net list, modules and workspace physical data);*
> *BEGIN*
> > *REPEAT*
> > > $A' = A;$
> > > $B' = B;$
> > > *FOR* $i = 1$ *TO* $n/2$ *DO BEGIN*
> > > > *Compute D values for all* $u \in A'$ *and* $v \in B'$;
> > > > *Find* u_i *and* v_i *that maximize* $C_{u_i v_i} = D_{u_i} + D_{v_i} - \delta_{u_i v_i}$;
> > > > *Move* u_i *to B and* v_i *to A;*
> > > > *Remove* u_i *and* v_i *from further consideration in this loop;*
> > > *END;* {*FOR* loop is a pass}
> > > *Find k that maximizes* $C_{max} = \sum_{i=1}^{k} C_{u_i v_i}$;
> > > *IF* $C_{max} > 0$, *THEN exchange* $u_1, ..., u_k$ *with* $v_1, ..., v_k$
> > *UNTIL* $C_{max} = 0$;
> *END.*

This procedure only indicates the minimizing process in the first partition of the breadth-first partitioning. When every partitioned subset has a small number of modules and the cost reduction is not equal to 0, a new breadth-first partitioning is repeated. Therefore, the objective of min-cut is to minimize the total number of external cuts (or external edges) generated by the breadth-first partitioning, as expressed by Equation 4.14. Associated with the reduction of the number of cuts, the average length of all external edges, $\bar{l}$, is also usually reduced because modules in the same net tend to be moved together in breath-first partitioning. According to Equation 4.15, the objective of min-cut approximates the minimum total length of minimum spanning trees.

If depth-first partitioning is employed, modules moved together in the previous partioning steps may be scattered in the subsequent partitioning steps because the number of the modules in each subset in the subsequent partitioning steps is larger than that in the previous partitioning steps. Therefore, l may not be reduced by decreasing the number of cuts, and the min-cut process using depth-first partitioning may not be convergent. In order to improve the placement with iterative partitioning, a multi-way partition [YEH91] and Steiner trees have been employed [WON89, MAY90].

Example: There are five selections in a min-cut loop in Figure 4.3. The cost reduction obtained from the exchanges in all selections are 2, 1, -1, 2 and -2, respectively. In this case, the local minimum occurs in selection 3. If the exchange is terminated in selection 2, the total cost is reduced by 3. If the exchange in selection 3 and the subsequent exchange in selection 4 are accepted, then the total cost reduction is 4.

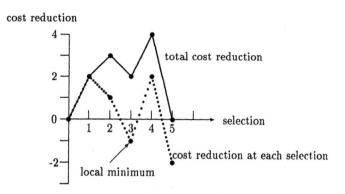

Figure 4.3. Cost reductions and a local minimum

◊

4.4   Approximating Minimum Steiner Trees

By simulating the movement of a node (module) of a tree (net), Y. T. Wong and M. Pecht [WON89] proposed an iso-distance error graph ($IDEG$), which contains connection errors generated by replacing a minimum rectilinear Steiner tree (MRST) with its equivalent non-MRST, in order to estimate how well various tree types function as the MRST in the placement, and construct a computationally efficient tree family, called a row-based tree family, by reducing connection errors in the $IDEG$. Then, based on the analysis of the error distribution on the $IDEG$, the row-based tree and the minimum spanning tree are compared in terms of their approximation to the MRST.

4.4.1   The iso-distance error graph

The total length of tree type X in the mth move of an iterative placement can be expressed as:

$$
\begin{aligned}
L_{X_m} &= L_{X_0} - \sum_{i=1}^{m} \Delta L_{X_i} \\
&= L_{X_0} - \sum_{i=1}^{m}(L_{X_i} - L_{X_{i-1}})
\end{aligned}
\tag{4.51}
$$

where L_{X_0} is the total tree length for tree X in the initial placement. Tree type X is said to be cost equivalent to the MRST in an iterative placement, if

$$
\Delta L_{XS} = \Delta L_X - \Delta L_S = 0
\tag{4.52}
$$

for each move, where S signifies the MRST.

One of the nodes in a tree can be designated as a moving node that can be moved to a new slot. The sub-tree without the moving node is called an associated tree. The tree length increase determined by an edge connecting the moving node to the associated tree is called the trace distance, where the edge is determined by the tree type. An iso-distance trace is generated by moving a node of a tree in slots around its associated tree without changing its trace distance. An iso-distance trace with distance n is called trace n of the tree type and denoted as T_n. When the placement of a moving node does not change the length of the associated tree, the set of these slots is defined as trace 0. Because there are many alternative trees for the same tree type in the same node configuration, trace 0 includes all slots occupied by all the associated trees for the same (fixed) node configuration.

The iso-distance graph (IDG) of a tree is a graph which consists of all iso-distance traces within the workspace. Because there is only one distance from an arbitrary slot to its associated tree, *no trace intersects with any other trace in an IDG*.

Example: Figures 4.4 and 4.5 show $IDGs$ for an MRST and a minimum spanning tree with five nodes, in which four nodes are shown and another can be moved to an arbitrary slot of the workspace. The tree connecting the shown nodes is called an associated tree. An edge length needed to connect the moving node and the associated tree is determined by the tree type selected.

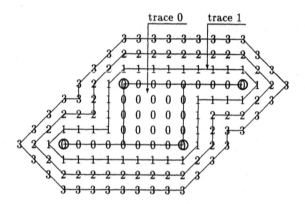

Figure 4.4. The IDG of a five node MRST

- The moving node can be placed in an arbitrary slot
- Trace 0 includes the complete area of "0s"
- There is no intersection between different traces
- T_n ($n \neq 0$) must be a closed line for a Steiner tree

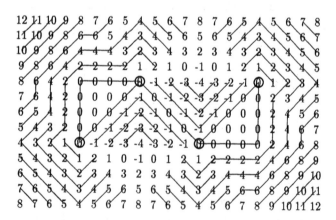

Figure 4.5. The IDG of the five node minimum spanning tree
(A negative number specifies a decrease in the tree
length when the moving node is placed in that slot)

In Figure 4.4, an associated tree consists of two horizontal edges and one vertical edge, and different associated trees can be obtained by horizontally moving the vertical edge. Therefore, trace 0 includes slots occupied by the two horizontal edges and all slots swept by moving the vertical edge. A slot labeled with 2 means connecting the moving node in this slot with the associated tree will increase the length of the associated tree by two units. Because of the node-to-node connecting property of the minimum spanning tree, properly locating the moving node will reduce the redundant edge in an associated tree for a minimum spanning tree. If the moving node is in the slot labeled -4, three shown nodes will be directly connected to the moving node, and the redundant edge with four units in the associated tree can be removed.

The line connecting slots with the same number is a trace. There are four traces in the IDG (see Figure 4.4). No trace can be intersected with other traces (see Figures 4.4 and 4.5). Except for trace 0, each trace in an IDG for an MRST in an open workspace is always closed, and trace 0 is enclosed by trace 1, trace 1 is enclosed by trace 2, and so forth, until the boundary is met.

◇

Let an IDG for an MRST be IDG_S and an IDG of another tree structure that has the same node configuration be IDG_X. Then the iso-distance error graph ($IDEG$) for IDG_S and IDG_X is defined as:

$$IDEG = \{\delta(x) = v(x) - u(x),\ T_i|\ u(x) \in IDG_S,\ v(x) \in IDG_X,$$
$$x \in T_i \in IDG_S,\ i = 1, ..., n\} \tag{4.53}$$

where x is the slot variable, u and v are the trace numbers for IDG_S and IDG_X in slot x, and $\delta(x)$ is the connection error associated with slot x.

Example: Figure 4.6 shows an example, $IDEG$, generated by subtracting the IDG for the the MRST in Figure 4.4 from the IDG for the minimum spanning tree in Figure 4.5, where the traces are exactly those shown in Figure 4.4. The number in each slot of the $IDEG$ is the difference between two numbers labeled in the corresponding slots in the IDE_S and IDG_X shown in Figures 4.4 and 4.5, respectively. Corresponding to the slot labeled with -4, shown in Figure 4.5, the slot in Figure 4.4 is labeled with 0. Therefore, the number in the corresponding slot in Figure 4.6 is $-4 - 0 = -4$.

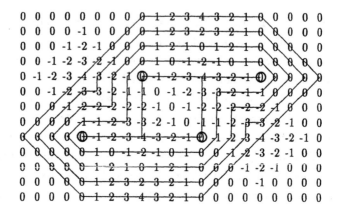

Figure 4.6. The $IDEG$ of the five node minimum spanning tree

(The increase of the minimum spanning tree length is greater than that of the MRST length for the moving node in a slot with positive connection error)

◇

4.4.2   The connection errors

To determine how well a tree configuration approximates the MRST, the cost difference produced by moving a node in an $IDEG$ from slot x to y is:

$$\Delta L_{XS} = \sum_{j=1}^{c} \Delta l_{XS_j}$$
$$= \sum_{j=1}^{c} (\Delta l_{X_j} - \Delta l_{S_j})$$
$$= \sum_{j=1}^{c} [\delta_j(x) - \delta_j(y)] \qquad (4.54)$$

where the subscript j refers to the jth of a total of c trees intersected to the moving node, Δl_{X_j} is the length change of the jth tree connected as a non-MRST, and Δl_{S_j} is the length change of the jth tree connected as an MRST. Given associated trees connecting the same node configuration for the MRST and tree X, Equation 4.41 can be written as:

$$\Delta l_S = \Delta l_X - [\delta(x) - \delta(y)] \qquad (4.55)$$

Reducing Δl_X from the tree length is equivalent to reducing Δl_S from the MRST length. If $\Delta l_X = n$ and $x \in T_n$, then

$$\Delta l_S = n - [\delta(x) - \delta(y)] \qquad (4.56)$$

where $y \in T_m(m = \Delta l_S)$. The MRST length is reduced $n - [\delta(x) - \delta(y)]$ units while the X tree length is reduced n units by moving a node from $x \in T_n$.

To simplify the placement problem defined in a space to a problem defined on an axis, a connection error, ξ_n of trace n for the $IDEG$, is defined as:

$$\xi_n = \frac{1}{M_n} \sum_{i=1}^{M_n} \delta(x_{ni}) \qquad (4.57)$$

where M_n is the number of slots in trace n.

Example: Figure 4.7 shows the $IDEG$ of a three node minimum spanning tree where there are twenty four slots on trace 1. Summing the value of the connection errors and dividing by the number of slots gives $\xi_1 = 20/24 = 0.83$. Similarly, on trace 0, $\xi_0 = 0$. On the average, the MRST length is reduced $1 - (0.83 - 0) = 0.17$ unit if the minimum spanning tree is used, while the MRST length is reduced one unit when the moving node is moved from trace 1 to trace 0 on the $IDEG$ of the tree configuration shown in Figure 4.7. Similarly, a negative ξ_1 can be obtained from Figure 4.6: $\xi_0 = -1.58$ and $\xi_1 = -0.76$. The MRST length is reduced $1 - [-0.76 - (-1.58)] = 0.18$ unit if the minimum spanning tree is used, while the MRST length is reduced one unit when the moving node is moved from trace 1 to trace 0 on the $IDEG$ of the tree configuration shown in Figure 4.6.

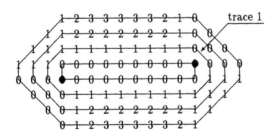

Figure 4.7. The $IDEG$ of a three node minimum spanning tree

◇

Because minimizing the length of the edge connecting the moving node is equivalent to reducing the tree length (i.e., moving the node from a higher trace to a lower trace and, if possible, trace 0), It is appropriate to define the average connection error, α_n, produced by moving a node a single-unit trace distance within trace n:

$$\alpha_n = (\xi_n - \xi_0)/n \qquad (n \neq 0) \qquad (4.58)$$

Example: There are thirty-three slots in trace 0, thirty-four slots in trace 1, and thirty-six slots in trace 3 of the $IDEG$ in Figure 4.6. Because the sum of errors is -52 in trace 0, -26 in trace 1, and -4 in trace 2, $\xi_0 = -1.58$, $\xi_1 = -0.76$, $\xi_2 = -0.11$, $\alpha_1 = 0.82$, and $\alpha_2 = 0.74$. On average, if the minimum spanning tree is used, the MRST length is reduced $1 - 0.82 = 0.18$ unit when a node is moved from trace 1 to trace 0, and is reduced $1 - 0.74 = 0.26$ unit per unit movement of a node that is moved from trace 2 to trace 0 on the $IDEG$ in Figure 4.6.
◇

4.4.3   The error index

From α_n, the average connection error of an $IDEG$ produced by moving a node from any trace to its adjacent lower trace in an $IDEG$, in which there are N_t traces and a total of M_{total} slots, which fit into the given workspace, is defined as:

$$
\begin{aligned}
E &= \sum_{n=1}^{N_t} \alpha_n M_n / (M_{total} - M_0) \\
&= [1/(M_{total} - M_0)] \sum_{n=1}^{N_t} [M_n(\xi_n - \xi_0)/n] \\
&= [1/(M_{total} - M_0)] \sum_{n=1}^{N_t} [\sum_{i=1}^{M_n} \delta(x_{ni}) - (M_n/M_0) \sum_{i=1}^{M_0} \delta(x_{0i})]/n \quad (4.59)
\end{aligned}
$$

Example: In the $IDEG$ shown in Figure 4.7, with trace 3 as the boundary of the workspace, $N_t = 3$, $M_0 = 20$, $M_1 = 24$, $M_2 = 28$, $M_3 = 32$, $M_{total} = \sum_{i=0}^{3} M_i = 104$, $\xi_0 = 0$, $\xi_1 = 20/M_1 = 20/24 = 0.83$, $\xi_2 = 36/M_2 = 36/28 = 1.29$, $\xi_3 = 48/M_3 = 48/32 = 1.5$ and $E = (\sum_{n=1}^{3} M_n \xi_n/n)/(M_{total} - M_0) = (20 + 36/2 + 48/3)/(104 - 20) = 0.64$. The error index, $E = 0.64$, implies that if the minimum spanning tree is used, the MRST length is reduced $1 - 0.64 = 0.36$ unit per unit movement of a node from a higher trace to a lower trace within trace 3 on the $IDEG$ in Figure 4.7, because $E = 0.64$ is computed from trace 0 to trace 3.
◇

Because not all trees will form an $IDEG$ with a non-zero E, those trees must be neglected in the computation of E. For example, a two-node spanning tree always functions as a two node Steiner tree. The average connection error for a workspace consisting of multiple trees is the average of all Es ($E \neq 0$) of $IDEGs$

of tree configurations within the workspace. Denoting the error index of tree type X for a workspace as E_X gives

$$E_X = (1/N_A) \sum_{i=1}^{N_W} [(1/N_i) \sum_{j=1}^{N_i} E_{ij}] \tag{4.60}$$

where N_W is the number of trees which can form the $IDEG$ with a non-zero E, N_A is the total number of trees, N_i is the number of configurations of tree i within the workspace, and E_{ij} is the average error of the $IDEG$ of the jth configuration of the ith tree.

The stopping criterion of placement is to maximize the accumulation of the cost Δl_X when tree type X is used. Since $\Delta l_X \mid_{\delta(x)>\delta(y)} > \Delta l_X \mid_{\delta(x)<\delta(y)}$ under the same Δl_S (see Equation 4.55), a node is moved from a slot with a large connection error to a slot with small connection error. Similarly, when maximizing the accumulation of Δl_X, $E_X < 0$ means nodes are moved from trace 0 to a higher trace in placement, and nodes in a tree will diverge from trace 0 to a higher trace by $E_X < 0$, because the connection errors in trace 0 can be considered to be 0 and the connection errors in a higher trace can be considered to be $- \mid E_X \mid$. The tree type with $E_X \ll 0$ cannot well approximate the MRST in placement.

Similarly, $E_X > 0$ means that nodes tends to be moved from a higher trace to trace 0. According to Equation 4.56, as a result of maximizing the accumulation of Δl_X, the MRST length is reduced $1 - E_X$ per unit movement from a trace with a greater average connection error to a trace with a lower average connection error. The tree type with a small E_x can well approximate the MRST in the placement. Therefore, the effects produced by both $E_X > 0$ and $E_X < 0$ are similar, and the placement for a non-MRST type can be understood as a process of reducing the total MRST length, with accumulating connection errors.

Because the calculation of E_X is complex, a representative error, E_r, is defined as $E_r \equiv \mid E_{m_1 n_1} + E_{m_2 n_2} + ... + E_{m_t n_t} \mid / t \approx \mid (\sum_{i=1}^{N_W} \sum_{j=1}^{N_i} E_{ij}) / (N_A N_i) \mid$ where t is the number of the selected $IDEGs$ and $t \ll \sum_{i=1}^{N_W} N_i$ (See section 4.4.4 for choosing the selected $IDEGs$). Using E_r, Equation 4.60 can be approximately expressed as:

$$E_X = (N_W / N_A) E_r \tag{4.61}$$

Using E_X, the relation between ΔL_{XS} and E_X is given as:

$$\sum_{i=1}^{m} \Delta L_{XS} = \sum_{i=1}^{m} (\Delta L_X - \Delta L_S)$$

$$= (1 + E_X)(L_{X_I} - L_{X_F}) - (L_{X_I} - L_{X_F})$$
$$= E_X(L_{X_I} - L_{X_F}) \qquad (4.62)$$

where s is the number of steps in a placement process, and L_{X_I} and L_{X_F} are the average total MRST length in the initial state and in the final state, respectively, when tree type X is used. Note that L_{X_I} is not equal to L_{X_0}, which is the total length of X-type trees. Equation 4.62 shows that the minimization of the cost difference can be transformed to the minimization of E_X. Similarly, the average reduction of the total MRST length in using tree type X can be computed by:

$$L_{X_I} - L_{X_F} = (1 - E_X)(L_{S_I} - L_{S_F}) \qquad (4.63)$$

where L_{S_I} and L_{S_F} are the average total MRST length in the initial state and in the final state, respectively, when the MRST is used. Considering $L_{X_I} = L_{S_I}$ and denoting $L_I = L_{S_I}$, Equation 4.63 becomes:

$$L_{X_F} = L_{S_F} + E_X(L_I - L_{S_F}) \qquad (4.64)$$

If $L_{X_{CF}}$ is the total MRST length in the final state of an arbitrary case when tree type X is used, then

$$L_{S_F} \leq L_{X_{CF}} \leq L_{S_F} + 2E_X(L_I - L_{S_F}) \qquad (4.65)$$

Therefore, $L_{X_{CF}}$ is changed in the range from L_{S_F} to $L_{S_F} + 2E_X(L_I - L_{S_F})$, and the change range of $L_{X_{CF}}$, $2E_X(L_I - L_{S_F})$ is determined by the error index of tree type X for a workspace. As E_X is increased, the change range of $L_{X_{CF}}$ is greater and the final placement becomes more sensitive to the changes of the initial states. When another tree type is used in the same workspace, the relationship of the final results and the error indices of two tree types are given as:

$$(L_I - L_{X_F})/(L_I - L_{Y_F}) = (1 - E_X)/(1 - E_Y) \qquad (4.66)$$

Using Equation 4.66, the result obtained by tree type Y can be predicted by the result obtained by tree type X. Setting $Y = S$ and $E_Y = 0$, the minimum total MRST length can be predicted by Equation 4.66, if L_{X_F} is obtained from minimizing the total length of the tree type X.

4.4.4 Characteristic *IDEGs*

In order to estimate which is the best for approximating the MRST among different tree types, a simple method of determining E_r is needed, because of the computational complexity of the error index E_X. Therefore, the property of *IDEG* requires further discussion.

The condition that exists at least one slot, $y \in T_{n-1}$ in an *IDG*, such that the distance between $x \in T_n$ and y is 1 is called the single-edge condition of x. This indicates that neither multiple edges nor a broken edge can exist between slots x and y. All slots that are not in trace 0 of an *IDG* for an MRST satisfy this condition.

Theorem 4.1. (Closed error area theorem)

If $\delta(x_{ni}) \equiv 0$ $(i = 1, ..., M_n)$ in trace n and slots in trace $m(m > n)$ satisfy the single edge condition in an *IDG* for a non-MRST, the connection errors must be enclosed by trace n.

Proof: Because $\delta(x_{ni}) = 0$ $(i = 1, ..., M_n)$ where $\delta(x_{ni})$ is the connection error in the x_{ni}-th slot of trace n, $S_n \subset N_n$ where S_n is the set of slots of trace n on the *IDG* for the MRST and N_n is the set of slots of trace n on the *IDG* for the non-MRST. Assume that at least one slot $l_n \in (N_n \cap S_{n+1})$ exists. According to the single-edge condition, a slot $l_{n-1} \in N_{n-1}$ must then exist, and the distance between l_n and l_{n-1} is 1. Because $l_n \in S_{n+1}$ and an arbitrary node with distance 1 from S_{n+1} is an element of S_n or S_{n+1}, $l_{n-1} \in S_n$ or $l_{n-1} \in S_{n+1}$. Because $S_n \subset N_n$, $l_{n-1} \in N_n$ if $l_{n-1} \in S_n$. This conflicts with $l_{n-1} \in N_{n-1}$, because l_{n-1} is an intersection of trace n and trace $n - 1$. Therefore, the property of the *IDG* is violated. If $l_{n-1} \in S_{n+1}$, at least one slot $l_{n-2} \in S_n \subset N_n$ or $l_{n-2} \in S_{n+1}$ exists. $l_{n-2} \in N_{n-2}$ is impossible because it conflicts with $l_{n-2} \subset N_n$. If $l_{n-2} \in S_{n+1}$, at least one slot $l_{n-1} \in S_n \subset N_n$ exists. The property of the *IDG* is also violated by $l_{n-1} \in S_{n+1}$. Hence, $l_n \in (N_n \cap S_{n+1})$ is impossible, and $N_n \cap S_{n+1} = \emptyset$. Assuming $l_{n+2} \in (N_{n+2} \cap S_{n+1})$, then $l_{n+1} \in S_n$ must exist. Because $S_n \subset N_n$, $l_{n+1} \in N_n$. Hence, $l_{n+2} \in (N_{n+2} \cap S_{n+1})$ is impossible, and $N_{n+2} \cap S_{n+1} = \emptyset$. Therefore, $S_{n+1} \subset N_{n+1}$ and $\delta(x_{(n+1)i}) \equiv 0$ $(i = 1, ..., M_{n+1})$. Substituting n with k, $\delta(x_{(k+1)i}) \equiv 0$ $(i = 1, ..., M_{k+1})$. Setting $k = n, ..., \infty$, $\delta(x_{ki}) \equiv 0$ $(k = n, ..., \infty;$ $i = 1, ..., M_k)$. $\square$

Corollary 4.1.1. If $\delta(x_{1i}) \equiv 0$ $(i = 1, ..., M_1)$, and if slots in trace 0 satisfy the single edge-condition in an IDG for a non-MRST, $\delta(x_{0i}) \equiv 0$ $(i = 1, ..., M_0)$.

According to the closed-error-area theorem and Equations 4.46 and 4.47, if $\delta(x_{1i}) \equiv 0$ $(i = 1, ..., M_1)$, then $E_X = 0$. The more connection errors in trace 1, the more redundant or broken interconnections produced by the interconnection rule of a tree type. The maximum redundant or broken edge which takes place in trace 1 will result in a maximum α_1 when $\xi_0 = 0$. By approximating the condition $E_X = 0$, only if the connection errors in trace 1 are minimized, will all connection errors in an $IDEG$ be minimized.

The $IDEG$ of the tree configuration of a tree type that gives the maximum E in a specific space is called the characteristic $IDEG$ of the tree type in that space. For convenience, the characteristic $IDEG$ can be approximately expressed by the $IDEG$ which gives the maximum ξ_1 in a specific number of slots in trace 1 when $\xi_0 = 0$. By minimizing ξ_1 of the characteristic $IDEG$ in a specific workspace, a tree type which approximately functions as the MRST in that workspace can be obtained.

It can be proven that the minimum number of slots which is located in the trace 1 and is adjacent to the nodes of an associated tree is 4 in an arbitrary $IDEG$ for the type of minimum spanning tree (see Figure 4.7). Therefore, the $IDEG$ for the 3 node minimum spanning tree shown in Figure 4.7 is the characteristic $IDEG$ for an unbounded workspace. The connection error, α_1, of the characteristic $IDEG$ is computed by:

$$\alpha_1 = \xi_1 = 1 - 4/M_1 \qquad (4.67)$$

where M_1 is the number of slots of trace 1. The connection error, α_1, of the characteristic $IDEG$ of the minimum spanning tree will be used to compare with the connection errors of other types of trees to estimate which tree type is the best in approximating the MRST.

Example: As shown in Figure 4.7, except for four slots in trace 1, that are neighboring to two nodes of the associated tree, the interconnection between a node in trace 1 and a slot in trace 0, following the connection rule of the minimum spanning tree, are redundant. In Figure 4.7, $\alpha_1 = \xi_1 = 1 - 4/24 = 0.83$.

◇

4.4.5 Approximations with row-based trees

To show how the developed theory can be used to create tree structures for approximating the MRST in placement, a new tree family called the row-based tree is constructed with the assumption that all nodes in a row of a tree are connected and adjacent rows are connected by a shortest distance. The length, L, of a row-based tree with n rows is computed by the equation:

$$L = \sum_{i=1}^{n}(l_i - f_i) + k_c \sum_{i=1}^{n-1}(r_{i+1} - r_i) + \sum_{i=1}^{n-1} h_i \qquad (4.68)$$

where f_i is the column coordinate of the first node in row i, l_i is the column coordinate of the last node in row i, r_i is the ith row coordinate, k_c is the ratio of the column unit and row unit used with gridded spacing when the column and row units are not equal (here $k_c = 1$ is chosen for simplicity), and h_i is the horizontal overlap between row i and row $i + 1$. The horizontal overlap, λ, between row i and row $i + inc$, is computed by the function:

FUNCTION $\lambda(f_i, l_i, f_{i+inc}, l_{i+inc})$;

BEGIN

   IF $f_{i+inc} > l_i$ *THEN* $\lambda = f_{i+inc} - l_i$

   ELSE IF $f_i > l_{i+inc}$ *THEN* $\lambda = f_i - l_{i+inc}$

   ELSE $\lambda = 0$; {There is an horizontal overlap between rows i and $i + inc$}
END.

The horizontal overlap, λ, is also the vertical projections of row i on row $i + inc$.

   The characteristic $IDEG$ of the row-based tree without correction can be determined by the redundancy produced from Equation 4.68, this $IDEG$ is shown in Figure 4.8. Three nodes are required to form the characteristic $IDEG$ and α_1 of the characteristic $IDEG$ is

$$\begin{aligned}
\alpha_1 = \xi_1 &= [1 + (M_1 - 6)/2][(M_1 - 6)/2]/M_1 \\
&= (M_1 - 4)(M_1 - 6)/(8M_1) \qquad (4.69)
\end{aligned}$$

A comparison of Equation 4.69 with Equation 4.67 reveals that α_1 of the characteristic $IDEG$ of the row-based tree without correction is greater then α_1 of the characteristic $IDEG$ of the minimum spanning tree when $M_1 > 14$. Thus, the minimum spanning tree works better than the row-based tree without correction in approximating the MRST in a workspace, in which there exists at least one *IDEG* with more than 14 slots in trace 1.

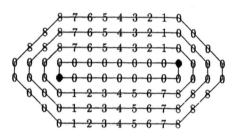

Figure 4.8. The characteristic *IDEG* of the row-based tree without correction

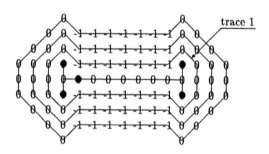

Figure 4.9. The characteristic *IDEG* of the row-based tree
with corrected terminal connections

In a row-based tree, the first and the last node of a row may not be connected by the shortest distance, and thus $\xi_1 \neq 0$ (see Figure 4.8). For example, if adjacent rows overlap horizontally and the distance between the first and the second nodes is greater than the distance between the adjacent rows, connecting all nodes in a row will increase connection error. If the moving node is the first or last node in a row, and if it is connected with the node in the same row or another row by the shortest distance, the connection error and the slots labeled with non-zero connection errors will be reduced. By comparing the distance between the first (last) node and the second (second to last) node and the distance between the first (last) node with other rows, the shortest connection between a terminal node and its associated tree can be determined. After correcting the terminal connections, connection errors are confined in a vertical band area, which is not wider than the maximum width of trace 0 (see Figure 4.9). The function $v(f_i)$, which is used to compute the trace number of the first node, is expressed as:

> FUNCTION $v(f_i)$;
>
> BEGIN
>
>   IF $(l_{i-1} \leq f_i \leq f_{i+1})$ OR $(l_{i+1} \leq f_i \leq f_{i-1})$
>
>   OR $[(l_{i-1} \leq f_i)$ AND $(l_{i+1} \leq f_i)]$ THEN $v = 0$
>
>   ELSE IF $(p_i \leq f_{i-1})$ AND $(p_i \leq f_{i+1})$ THEN $v = p_i - f_i$
>
>   ELSE $v = min(p_i - f_i, d_{i-1}, d_{i+1}, h_c)$
>
> END.

In this function, p_i is the coordinate of the second node in the i-th row, d_x is the distance between f_i and the row x, h_c is the shortest distance between f_i and the vertical connection of the rows $i - 1$ and $i + 1$, without regard to the existence of row i. Similarly, the function $v(l_i)$, which is used to compute the trace number of the last node in the i-th row, can be obtained. If $x \neq f_i$ and $x \neq l_i$, then $v(x) = 0$. Because the change in the length of the row-based tree at the i-th step is

$$\Delta L_{R_i} = \sum_{j=1}^{c} (v_{i-1,j} - v_{i,j}) \tag{4.70}$$

where c is the number of trees that connect the moving node, and $v_{i,j}$ is the trace number of the ith step in the IDG for the row-based tree identified as j. Equation 4.70 can be used to be the cost function without regard to the weighting and correction functions, and therefore the cost can be computed by functions $v(f_i)$, $v(l_i)$ and $v(x)$ $(x \neq f_i$ and $x \neq l_i)$. The row-based tree on which the row terminal connections are computed by functions $v(f_i)$, $v(l_i)$ and $v(x)$ $(x \neq f_i$ and $x \neq l_i)$ is called a row-based tree with corrected terminal connections.

Because the computation of $v(f_i)$ is only related to three rows, the time complexity for computing the length of a row-based tree with corrected terminal connections is bounded by the number of rows, n_{row}, in the workspace, or the average number of nodes, n_{node}. Therefore, the time for computing the length of a row-based tree with corrected terminal connections is linear to $max(n_{row}, n_{node})$.

Assuming the length of trees identified as 1 to a become shorter, the length of trees identified as $(a + 1)$ to b remains the same, and the length of trees identified as $(b + 1)$ to c becomes longer in swapping modules A and B, Equation 4.70 can be rewritten as:

$$\Delta L_{X_i}(A, B) = \sum_{j=1}^{a}(v_{i-1,j} - v_{i,j}) + \sum_{j=b+1}^{c}(v_{i-1,j} - v_{i,j}) \qquad (4.71)$$

Before a terminal node is selected as a moving node, it must be connected with its adjacent node in the same row. As soon as the node, f_i, is selected, it is connected with the adjacent row if $(r_i - r_{i+inc}) + \lambda(f_i, f_i, f_{i+inc}, l_{i+inc}) < f_i - p_i$. This is different from a traditional tree in which the connection is independent of whether a node is selected as a moving node. The change of tree length before and after selecting a moving node is equivalent to resetting L_{X_0} in Equation 4.51. If the terminal connections of each column are also corrected in the row-based tree, the row-column-based tree is constructed. The characteristic $IDEG$ of this tree type has a closed trace 0, and connection errors only exist in parts of other traces. Therefore, ξ_1 of the characteristic $IDEG$ of this tree type is much smaller than that of the row-based tree with corrected (row) terminal connections.

Because all connection errors are confined in a band area not wider than the maximum width of trace 0, the characteristic $IDEG$ of the row-based tree with corrected terminal connections is shown in Figure 4.9. It takes at least six nodes to form the characteristic $IDEG$ in an unbounded workspace, and α_1 is

$$\alpha_1 = \xi_1 = 1 - 10/M_1 \qquad (4.72)$$

where M_1 is the number of slots in trace 1. A comparison of this result with that given by Equation 4.67 reveals that α_1 of the characteristic $IDEG$ of the minimum spanning tree is greater than that of the $IDEG$ of the row-based tree with corrected terminal connections.

4.4.6   Testing and discussion

To show how the error index can be approximately computed from the characteristic $IDEG$, and how some specific non-MRST types approximate the MRST, a twenty-four-slot, 8×3, workspace was selected, in which there were twenty-seven nets, twenty-two modules, and two unoccupied slots. All modules were assumed to be 1 unit $\times$ 1.8 unit (see Figure 4.10), because this size is similar to that of a DIP (dual-in-line package). Because the net list, the module size, and the workspace were simple, the influence of an iterative process could be ignored and the final placement was affected only by the cost function corresponding to tree type.

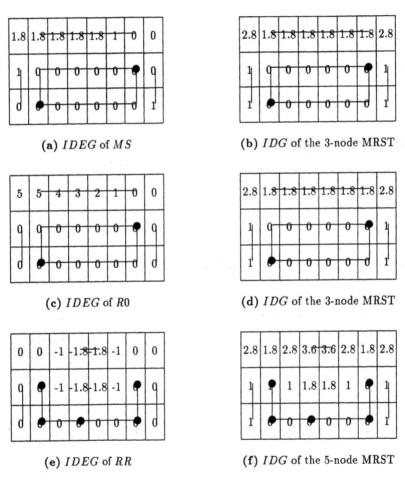

Figure 4.10. Characteristic *IDEGs* for computing
error indices of a 8 × 3 workspace

The maximum accumulated-cost-reduction iterative technique was used, and exchanges which result in the maximum cost reduction, ΔL_{X_i}, were performed. In the present testing, X indicates the minimum spanning tree (**MS**), or the row-based tree without correction (**R0**), or the row-based tree with corrected terminal connections (**RR**). Figure 4.10 shows the placement configurations obtained by using the tree types **MS**, **R0** and **RR** for different initial placements. The total MRST length in each resulting configuration is shown in Table 4.2.

Table 4.2.   Testing placement results

Objective	Initial	MS	R0	RR
Total MRST length (Case 1)	179.6	173.6	159.8	165.4
Total MRST length (Case 2)	218.8	175.4	190.0	161.4
Total MRST length (Case 3)	192.8	167.4	185.2	161.2
Average length small	197.1	172.1	178.3	162.7

MS is the minimum spanning tree.
R0 is the row-based tree without correction.
RR is the row-based tree with corrected terminal connections.

To test the relationship given in Equation 4.66, characteristic $IDEGs$, in which the trace 0 within a half of the total slots of the workspace is located in the center of the workspace, are selected as representative $IDEGs$, and they are then used to compute the representative errors for the three tree types. As shown in Figure 4.10, there are twenty-four slots in the workspace, and half of the total slots is twelve. An area which occupied twelve slots and is located approximately in the center of the workspace is selected to locate trace 0 of the characteristic $IDEG$ of a tree type (see Figures 4.10), This $IDEG$ is then used to compute E_r in Equation 4.61. Because $N_i = 1$ in Equation 4.60, $\xi_0 = 0$ in Equation 4.59, and trace numbers are real number in the testing examples, E in Equation 4.59 is revised to compute E_r for real-number traces, $t_1, t_2, ..., t_n$:

$$E_r =| \sum_{n=1}^{N_t}(M_{t_n}\xi_{t_n}/t_n) | /(M_{total} - M_0) \qquad (4.73)$$

In Figure 4.10a, $N_A = 27$, $N_W = 14$, $M_{total} = 24$, $M_0 = 12$, $N_t = 3$, $t_1 = 1$, $t_2 = 1.8$, $t_3 = 2.8$, $M_1 = 4$, $M_{1.8} = 6$, $M_{2.8} = 2$, $\xi_1 = (1 + 0 + 0 + 1)/4 = 0.5$, $\xi_{1.8} = (4 \times 1.8 + 1)/6 = 1.37$, and $\xi_{2.8} = (1.8 + 0)/2 = 0.9$. Substituting the data into Equations 4.60 and 4.48 yields $E_r = 0.6$ and $E_X = 0.31$. The computation of the three error indices for the workspace with $N_A = 27$ and $M_{total} = 24$ is shown in Table 4.3, and the predicted total MRST length computed by Equation 4.66 is shown in Table 4.4.

Table 4.3. Computations of the average error indices

Tree Type (X)	MS	R0	RR
N_W	14	14	7
N_t	3	3	4
M_1	4	4	6
$M_{1.8}$	6	6	4
$M_{2.8}$	2	2	4
$M_{3.6}$	-	-	2
ξ_1	0.5	0	0.33
$\xi_{1.8}$	1.37	2.5	0.9
$\xi_{2.8}$	0.9	2.5	0.5
$\xi_{3.6}$	-	1.7	1.8
E_X	0.31	0.44	0.09

Table 4.4. Predicting the result of tree type Y by the result of tree type X

Tree Type X	MS	MS	RR	MS
Tree Type Y	RR	MRST	MS	R0
E_X	0.31	0.31	0.09	0.31
E_Y	0.09	0	0.31	0.44
L_I	197.1	197.1	197.1	197.1
L_{X_F}	172.1	172.1	162.7	172.1
$L^*_{Y_F}$	164.2	160.9	171.0	176.7
$L^{**}_{Y_F}$	162.7	-	172.1	178.3

$L^*_{Y_F}$ is the predicted value.
$L^{**}_{Y_F}$ is the testing value.

The error indices of tree types, **MS**, **R0** and **RR**, for the workspace are 0.31, 0.44 and 0.09, respectively (see Table 4.3). The final average total MRST length produced by the placement is $L^{**}_{(RR)_F} = 162.7$ when the tree type **RR** is used, and the final average total MRST length predicted by the final average total MRST length obtained by the tree type **MS** is $L^*_{(RR)_F} = 164.2$. Similarly, $L^{**}_{(R0)_F} = 178.3$ and $L^*_{(R0)_F} = 176.7$, when the final average total Steiner tree length obtained by tree type **MS** is used to predict the final average total Steiner tree length obtained by the tree type **R0**.

Because three nodes are required to form the representative $IDEG$ of the tree type of **MS** or **R0** (see Figures 4.10a through 4.10d), and at least six nodes are required to form the representative $IDEG$ of the tree type **RR** (see Figures 4.10e and 4.10f), the number of trees which can form the representative $IDEG$ of **MS** or **R0** is fourteen, and the number of trees which can form the representative $IDEG$ of the tree type **RR** is seven (see Table 4.3). The number of **MS**s which can form the representative $IDEG$ is more than the number of **RR**s which can form the representative $IDEG$. Furthermore, the representative error of the tree type **MS** is larger than that of the tree type **RR**. According to Equation 4.60, $E_{MS} \gg E_{RR}$. Therefore, the tree type **RR** is better than the tree type **MS**.

According to Equation 4.65, the change range of the MRST length in the final state obtained by using **R0** is $2E_X(L_I - L_{S_F}) = 2 \times 0.44(197.1 - 160.9) = 31.9$. Comparing to Table 4.2, the change range obtained by using **R0** in the test is $190.0 - 159.8 = 30.2$. For tree type **RR**, the change range $2E_X(L_I - L_{S_F})$ is 6.5, and the change range obtained by the test is 4.2. If E_X is less, the final placement is more independent of the initial placement. Therefore, the error index for a specific workspace dominates the sensitivity affecting the placement result through the change of the initial placement. Improving the tree type is critical in approximating the MRST in order to raise the placement quality.

4.5   References

[BLA85]   Blanks, J. P., "Near-Optimal Placement Using A Quadratic Objective Function," *Proc. of the 22rd Design Automation Conf.*, 1985, pp. 609-615.

[BRE72]   Breuer, M. A., *Design Automation of Digital Systems*, Prentice-Hall, Englewood Cliffs, N. J,. 1972.

[BRE77]   Breuer, M. A., "A Class of Min-Cut Placement Algorithm," *Proc. of 14th Design Automation Conf.*, 1977, pp. 284-290.

[CHE84]   Cheng, C. K. and Kuh, E., "Module Placement Based on Resistive Network Optimization," *IEEE Trans. on Computer-Aided Design*, vol. 3, no. 7, July. 1984, pp. 218-225.

[CIA75]   Ciampi, P. L., "A System for Solution of the Placement Problem," *Proc. of the 12th Design Automation Conf.*, 1975, pp. 317-323.

[COR79]   Corrigan, L. I.. "A Placement Capability Based on Partitioning," *Proc. of the 16th Design Automation Conf.*, 1979, pp. 406-413.

[DUN85]   Dunlop, A. E., and Kernighan, B. W., "A Procedure for Placement of Standard-cell VLSI Circuit," *IEEE Trans. on Computer-Aided Design*, vol. 4, no. 1, Jan. 1985, pp. 92-98.

[FID82]   Fiduccia, C. M., and Matteyses, R. M., "A Linear-Time Heuristic for Improving Network Partitions," *Proc. of 19th Design Automation Conf.*, 1982, pp. 175-181.

[FOR87]   Forbes, R., "Heuristic Acceleration of Force-Directed Placement," *Proc. of the 24th Design Automation Conf.*, 1987, pp. 735-740.

[GOT86]   Goto, S., and Matsuda, T., "Partitioning, Assignment and Placement," in *Layout Design and Verification*, edited by T. Ohtsuki, North-Holland, Amsterdam, 1986, pp. 55-97.

[GRO87]   Grover, L. K., "Standard Cell Placement Using Simulated Sintering," *Proc. of 24th Design Automation Conf.*, 1987, pp. 56-59.

[HAN66]   Hanan, M., "On Steiner's Problem with Rectilinear Distance," *SIAM J. of Appl. Math.*, vol. 14, no. 3, March, 1966. pp. 255-265.

[HAN72]   Hanan, M., and Kurtzberg, J. M., "Placement Techniques," in M.A. Breuer, ed., *Design Automation of Digital Systems, Volume one: Theory and techniques*, Prentice-Hall, Englewood Cliffs, New Jersey, 1972, ch. 5, pp. 213-282.

[HAN76]   Hanan, M., Wolff, P. K. Sr., and Agule, B. J., "Some Experimental Results on Placement Techniques," *Proc. of the 13th Design Automation Conf.*, 1976, pp. 214-224.

[HAR86]   Hartoog, M. R., "Analysis of Placement Procedure for VLSI Standard Cell Layout," *Proc. of the 23rd Design Automation Conf.*, 1986, pp. 314-319.

[HU85]   Hu, T. C., and Kuh, E. S., *VLSI Circuit Layout: Theory and Design*, *IEEE Press*, New York, 1986.

[IOS83]   Iosupovici, A., King, C., and Breuer, M. A., "A Model Interchange Placement Machine," *Proc. of the 19th Design Automation Conf.*, 1983, pp. 457-464.

[JOH89]   Johnson, D. S., Aragon, C. R., McGeoch, L. A., and Sechevon, C., "Optimization by Simulated Annealing: An Experimental Evaluation," *Operations Research*, vol. 37, no. 6, 1989, pp. 864-892.

[KER70]   Kernighan, B. W., and Lin, S., "An Efficient Heuristic Procedure for Partitioning Graphs," Bell Syst. Tech. J., vol. 49, no. 2, 1970, pp. 291-308.

[KHO77]   Khokhani, K. H., and Patel, A. M., "The Chip Layout Problem: a Placement Procedure for LSI," *Proc. of the 14th Design Automation Conf.*, 1977, pp. 219-297.

[KIR83]   Kirkpatrick, S., Gelatt, C. D., and Vecchi, M. P., "Optimization by Simulated Annealing," *Science*, vol. 220, no. 4598, 1983, pp. 671-680.

[KLE88]   Kleinhans, J. M., Sigl, G., and Johannes, F. M., "GORDIAN: A New Global Optimization/Rectangle Dissection Method for Cell Placement," *Proc. IEEE Int. Conf. on CAD*, 1988, pp. 506-509.

[KLI87]   Kling, R-M., and Banerjee, P., "A New Standard Cell Placement Package Using Simulated Evolution," *Proc. of the 24nd Design Automation Conf.*, 1987, pp. 60-66.

[KLI90]   Kling, R-M., and Banerjee, P., "Optimization by Simulated Evolution with Applications to Standard Cell Placement," *Proc. of the 24nd Design Automation Conf.*, 1990, pp. 20-25.

[KOZ83]   Kozawa, T., Terai, H., Ishii, T., Hayase, M., Miura, C., Ogawa, Y., Kishida, K., Yamada, N., and Ohno, Y., "Automatic Placement Algorithms for High Packing Density VLSI," *Proc. of the 20th Design Automation Conf.*, 1983, pp. 175-181.

[LAM88]   Lam, J., and Delosme, "Performance of a New Annealing Schedule," *Proc. 25th Design Automation Conf.*, June 1988, pp. 306-311.

[LAU79]   Lauther, U., "A Min-Cut Placement Algorithm for General Cell Assemblies Based on Graph Partitioning," *Proc. 16th Design Automation Conf.*, June 1979, pp. 1-10.

[LUN83]   Lundy, M., and Mees, A., "Convergence of the Annealing Algorithm," University of Cambridge, London, 1983.

[MAY90]   Mayrhofer, S., and Lauther, U., "Congestion-Driven Placement Using a New Multi-partitioning Heuristic," *Proc. IEEE Int. Conf. on CAD*, 1990, pp. 332-335.

[NAH85]   Nahar, S., Sahni, S., and Shragowitz, E., "Experiments with Simulated Annealing," *Proc. of the 22nd Design Automation Conf.*, 1985, pp. 784-752.

[NAH86]   Nahar, S., Sahni, S., and Shragowitz, E., "Simulated Annealing and Combinatorial Optimization," *Proc. of the 23rd Design Automation Conf.*, 1986, pp. 293-299.

[OHT86]   Ohtsuki, T., *Layout Design and Verification*, North-Holland, Amsterdam, 1986.

[PEC91]   Pecht, M., *Handbook of Electronic Package Design*, Marcel Dekker, New York, 1991.

[PRE86]   Preas, B. T., and Karger, P. G., "Automatic Placement: A Review of Current Techniques," *Proc. of the 23rd Design Automation Conf.*, 1986, pp. 622-629.

[QUI79]   Quinn Jr., N. R., and Breuer, M. A., "A Force Directed Component Placement Procedure for Printed Circuit Boards," *IEEE Trans. on Circuits and Systems*, vol. 26, no. 6, June 1979, pp. 377-387.

[ROM84]   Romeo, F., Sangiovanni-Vincentelli, A., and Sechen, C., "Research on Simulated Annealing at Berkeley," *Proc. IEEE Int. Conf. on CAD*, 1984, pp. 652-657.

[ROM85]   Romeo, F., and Sangiovanni-Vincentelli, A., "Probabilistic Hill Climbing Algorithms: Properties and Applications," *Proc. of the 1985 Chapel Hill Conf. on VLSI*, 1985, pp. 393-417.

[ROS90]   Rose, J., Klebsch, W., and Wolf, J., "Temperature Measurement and Equilibrium Dynamics of Simulated Annealing Placements," *IEEE Trans. on Computer Aided Design*, vol. 9, no. 3, March 1990, pp. 253-259

[SAA90]   Saab, Y. G., and Roa, V. B., "Stochastic Evolution: A Fast Effective Heuristic for Some Generic Layout Problems," *Proc. 27th Design Automation Conf.*, 1990, pp. 26-31.

[SCH72]   Schuler, D. M., and Ulrich, E. G., "Clustering and Linear Placement," *Proc. 9th Design Automation Conf.*, 1972, pp. 50-56.

[SEC88]   Sechen, C., *VLSI Placement and Global Routing Using Simulated Annealing*, Kluwer Academic Publishers, Boston, 1988.

[SIG91]   Sigl, G., Doll, K., and Johannes, F. M., *Analytical Placement: A Linear or Quadratic Objective?* *Proc. 28th Design Automation Conf.*, 1991, pp. 427-432.

[STE61]   Steinberg, L., "The Backboard Wiring Problem: a Placement Algorithm," *SIAM Review*, vol. 3, no. 1, January 1961, pp. 37-50.

[SUA88]   Suaris, P. R., and Kedem, G., "An Algorithm for Quadri-section and its Application to Standard Cell Placement," *IEEE Trans. on Circuits and Systems*, vol. CAS-35, no. 4, May, 1988, pp. 294-303.

[TSA91]   Tsay, R. S., and Kuh, E., "A Unified Approach to Partitioning and Placement," *IEEE Trans. on Circuits and Systems*, vol. CAS-38, no. 5, May, 1991, pp. 521-533.

[VEC83]   Vecchi, M., and Kirkpatrick, S., "Global Wiring By Simulated Annealing," *IEEE Trans. on Computer Aided Design*, vol. 2, no. 4, Oct. 1983, pp. 215-222.

[VIJ89]   Vijayan, G., "Min-Cost Partitioning on A Tree Structure and Application," *Proc. 26th Design Automation Conf.*, 1989, pp. 775-778.

[WHI84]   While, S., "Concepts of Scale in Simulated Annealing," *Proc. IEEE Int. Conf. on CAD*, Oct. 1984, pp. 646-651.

[WEI89]   Wei, Y-C., and Cheng, C-K., "Toward Efficient Hierarchical Designs by Ratio Cut Partitioning," *Proc. IEEE Int. Conf. on CAD*, 1989, pp. 298-301.

[WON89]   Wong, Y. T., and Pecht, M., "Approximating the Steiner Tree in the Placement Process", *ASME J. of Electronics Packaging*, Sept. 1989. pp. 228-235.

[YEH91]   Yeh, C-W., Cheng, C-K., and Lin, T-T. Y., "A General Purpose Multiple Way Partitioning Algorithm," *Proc. 28th Design Automation Conf.*, 1991, pp. 421-426.

Chapter 5

Placement for Reliability and Producibility

Michael D. Osterman and Michael Pecht

Placement techniques were initially developed to minimize the total weighted wire length necessary for complete interconnection. But with the increased demand for high quality and reliable performance over time, techniques to address placement for reliability and producibility also became necessary. The physical layout of electronic modules influences the reliability, manufacture, cost, and electrical performance of the product. Judiciously placing modules can both reduce the overall stresses that lead to failure, and provide ease and neatness of routability for enhancing yields.

The reliability of an electronic assembly is based in part on the failure mechanisms of its individual parts and interconnections. Failure mechanisms can be accelerated by temperature extremes, temperature changes and gradients, thermal-mechanical stresses, vibration, corrosion and electrical stresses. The goal of placement for reliability is to locate modules in such a manner as to reduce the stresses which can lead to failure, or to reduce the impact of the stresses on the potential failure mechanisms. Pecht [PEC91] and Hakim [HAK89] have compiled some of the common failure mechanisms of microelectronic modules. The placement of integrated circuits on a substrate for reliability has been studied by several researchers [MAY78, POR78, BUR78, WOL81, MAY81, ALT82, PEC87, OST89]. Mayer [MAY78] discussed the thermal design of avionics based on cost and reliability; Pecht, et al. [PEC87] examined reliability and routability tradeoffs; and Osterman and Pecht [OST89] developed a priority indexing scheme for placement of modules to maximize circuit card reliability.

Placement for producibility involves addressing the manufacturing and assembly processes, including testing and rework, by taking into account the equipment used, the process flow and human factors. The goal of placement for producibility is to locate modules to optimize the yield and to minimize the possibility of inducing a defect into the product.

5.1   Placement for Temperature Dependent Reliability

Placement for reliability requires an understanding of the actual operating conditions, under which the assembly will be used and the physical mechanisms involved in potential failures. For an assembly consisting of N independent failure opportunities for modules and interconnects, the total reliability, R, is defined by:

$$R = \prod_i^N R_i \qquad (5.1)$$

The reliability of an individual failure opportunity for any module or interconnect, R_i, is defined by:

$$R_i = e^{-\int_0^t h_i(S,\bar{x},t)dt} \qquad (5.2)$$

where $h_i(S,\bar{x},t)$ is the hazard rate based on the failure mechanisms; S is the stress caused by temperature, temperature change, vibration, and the environment in general; $\bar{x}$ represents the module's position on the substrate; and t is time. If the hazard rate of the individual module is

$$h_i = h_i(S,\bar{x})f(t) \qquad (5.3)$$

where $f(t)$ is the time relation and $h_i(S,\bar{x})$ is the stress dependent hazard rate term, the total reliability can be written as:

$$R(t) = e^{-\sum_i^N h_i \int_0^t f(t)dt} \qquad (5.4)$$

One approach to maximize reliability is to minimize $\sum h_i$. Another approach is to place modules on the workspace based on successively maximizing the minimum time to failure of the modules and interconnections.

This section develops placement objectives based on minimizing the total hazard rate, using temperature as an example stress. For practicality, both forced convection and conduction cooling technologies are examined, and an approach to placement for a complete substrate is presented.

5.1.1   Convection-cooling placement

This subsection establishes a theoretical basis for reliability placement on a forced-convection-cooled substrate. Forced-convection cooling is restricted to flow across the substrate with known fluid properties and a specified inlet temperature. A typical forced-convection cooling model is depicted in Figure 5.1. The temperature of individual electronic modules is modeled, using a thermal resistance path between each module and the working fluid. The working fluid temperature is assumed to be a function of the heat dissipated by the individual electronic modules through which the fluid has passed. Thus, the fluid temperature associated with any module at position s is approximated by:

$$T_f(s) = T_{inlet} + \sum_{i \in B(s)} \frac{q_i}{\dot{m} C_p} \tag{5.5}$$

where T_{inlet} is the inlet fluid temperature, $\dot{m}$ is the mass flow rate of the fluid, C_p is the specific heat of the fluid, $B(s)$ is the set of modules between the inlet and position s, and q_i is the heat dissipated by the ith module. The module temperature of the ith module on a row is

$$T_i(s) = T_f(s) + q_i R c f_i \tag{5.6}$$

where $R c f_i$ is the thermal resistance between the module and the working fluid. For thermal or power cycling, the cyclic temperature difference, ΔT_i, is assumed to be the difference between $T_i(s)$ and a specified base temperature:

$$\Delta T_i = T_i - T_{b_i} = T(s) + q_i R c f_i - T_{b_i} \tag{5.7}$$

where T_{b_i} is a specified base temperature associated with the module and ambient substrate conditions. The base temperature assumes that the entire substrate returns to an ambient temperature. This model provides a first-order prediction of temperature trends that occur in the substrate and the modules.

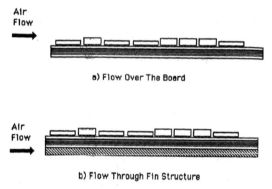

a) Flow Over The Board

b) Flow Through Fin Structure

Figure 5.1. Forced convection cooling models

For a row of modules as depicted in Figure 5.1, the total temperature dependent hazard rate is given by h_T. If two adjacent modules, i and $i + 1$, are switched, the new total temperature dependent hazard rate is given by h_T' (See Figure 5.2). The calculation of the module temperature and the module temperature difference affects only the interchanged modules. Thus, the change in the total temperature dependent hazard rate is

$$
\begin{aligned}
\Delta h_T &= h_T' - h_T \\
&= h_i(T_i^d, \Delta T_i^d) - h_i(T_i^u, \Delta T_i^u) \\
&\quad + h_{i+1}(T_{i+1}^u, \Delta T_{i+1}^u) - h_{i+1}(T_{i+1}^d, \Delta T_{i+1}^d)
\end{aligned}
\tag{5.8}
$$

where the superscripts u and d denote the upstream and downstream adjacent positions, respectively. Because the module temperature and the module temperature difference are both dependent on the associated fluid temperature, the individual temperature dependent hazard rates can be expressed in a Taylor series expansion based on either the upstream or downstream position, provided that the series is convergent. For example, $h_i(T_i^d, \Delta T_i^d)$ can be expressed as:

$$
h_i(T_i^d, \Delta T_i^d) = h_i(T_i^u, \Delta T_i^u) + \sum_{k=1}^{\infty} \frac{d^k h_i(T_i^u, \Delta T_i^u)}{dTf^k} \cdot \frac{(Tf_i^d - Tf_i^u)^k}{k!}
\tag{5.9}
$$

From Equation 5.5, it can be shown that fluid temperature differences resulting from the interchange of modules i and i+1 are

$$
Tf_i^d - Tf_i^u = \frac{q_{i+1}}{\dot{m} c_p}
\tag{5.10}
$$

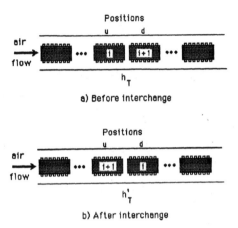

Figure 5.2. Module interchange for forced convection

and

$$T f_{i+1}^d - T f_{i+1}^u = \frac{q_i}{\dot{m} c_p} \tag{5.11}$$

Substituting the Taylor expansions for the downstream terms of modules i and $i+1$ in Equation 5.8, and multiplying through by $\dot{m} C_p / (q_i q_{i+1})$, yields

$$
\frac{\Delta h_T \dot{m} C_p}{q_i q_{i+1}} = \sum_{k=1}^{\infty} \frac{d^k h_i(T_i^u, \Delta T_i^u)}{dT f^k} \frac{1}{q_i k!} \left(\frac{q_{i+1}}{\dot{m} C_p} \right)^{k-1}
$$
$$
- \sum_{k=1}^{\infty} \frac{d^k h_{i+1}(T_{i+1}^u, \Delta T_{i+1}^u)}{dT f^k} \frac{1}{q_{i+1} k!} \left(\frac{q_i}{\dot{m} C_p} \right)^{k-1} \tag{5.12}
$$

According to Equation 5.12, the change in total hazard rate can be positive, negative, or zero. The primary objective is to position the modules in such a way as to ensure that the total hazard rate is minimum. If an interchange can result only in a positive change in the total hazard rate, then the modules are in their optimum positions. This assumption provides a rule for ordering modules on a given row. The condition for optimum reliability based on the module positions on a row is given by the following inequality:

$$
\sum_{k=1}^{\infty} \frac{d^k h_i(T_i^u, \Delta T_i^u)}{dT f^k} \frac{1}{q_i k!} \left(\frac{q_{i+1}}{\dot{m} C_p} \right)^{k-1}
$$
$$
\geq \sum_{k=1}^{\infty} \frac{d^k h_{i+1}(T_{i+1}^u, \Delta T_{i+1}^u)}{dT f^k} \frac{1}{q_{i+1} k!} \left(\frac{q_i}{\dot{m} C_p} \right)^{k-1} \tag{5.13}
$$

This inequality suggests a placement procedure based on a positioning metric for the convection-cooling case. The convection placement metric (CVPM) of any order, n, is defined as:

$$CVPM_i^n(u) = \sum_{k=1}^{n} \frac{d^k h_i(T_i^u, \Delta T_i^u)}{dTf^k} \frac{1}{q_i k_!} \left(\frac{q_a}{\dot{m}c_p}\right)^{k-1} \qquad (5.14)$$

where q_a is the heat dissipated by the module immediately downstream of module i. If the modules are placed so that the individual positioning metric decreases as the modules are located from the inlet end to the outlet end of the row, the prescribed placement configuration yields the minimum total hazard rate for a given row. Unfortunately, Equation 5.14 requires knowledge of the heat dissipation rate of the downstream module. To resolve this problem, a first-order approximation of the metric can be employed:

$$CVPM_i^1(u) = \frac{dh_i(T_i^u, \Delta T_i^u)}{dTf^k} \frac{1}{q_i} \qquad (5.15)$$

This metric is only dependent on the first derivative of the hazard rate, the fluid temperature, and the heat dissipation rate of the module.

An approximate but efficient placement procedure for minimizing the total hazard rate on a row consists of placing modules according to the convection placement metric (CVPM) defined by Equation 5.15. This may be accomplished through the following steps:

Step 1: Assume that modules can be placed at any position along the row, provided that module overlap does not occur. The placement positions are initially all open (i.e., all modules are initially unplaced) and the set A of placed modules is initially empty. Target placement positions are selected based on proximity to the inlet location. The first target position selected is the one close to the inlet end. Determine the resistance from module to fin, Rcf, for each module, using a typical resistance networking technique.

Step 2: Calculate the fluid temperature at the target location for every unplaced module. For convective cooling, the increased fluid temperature is given by Equation 5.5.

Step 3: Calculate the module temperature for every unplaced module, using Equation 5.6. Using the calculated module temperatures and the corresponding heat dissipation rates, determine the CVPM of every unplaced module, using Equation 5.15.

Step 4: Select the module with the highest CVPM value. Place this module in the target location and add it to A. If two or more modules have the same CVPM value, an arbitrary selection may be made.

Step 5: Select the next target position downstream and immediately adjacent to the previously placed module. Repeat steps 2 through 5 until all modules are placed. The force-convection placement procedure is illustrated in Figure 5.3.

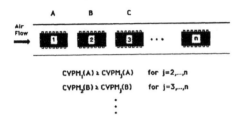

Figure 5.3. Forced convection placement rule

5.1.2 Conduction cooling placement

In this section, a reliability placement theory for conduction-cooled substrates is developed. Two examples are then presented.

If the modules are optimally placed, switching any two modules will result in an increased total hazard rate, h_T'. In particular, if modules i and $i + 1$ are switched as in Figure 5.4, the difference in the total hazard rate is

$$(h_T' - h_T) = \sum_{k=1}^{N}[h_k'(T_k') - h_k(T_k)] \qquad (5.16)$$

where T_k' is the temperature of the k-th module, which occurs as a result of interchanging modules i and $i + 1$; T_k is the temperature of k-th module for the original placement configuration; and N is the number of modules on the row. Because h_T is assumed to be the minimum total hazard rate, both sides of Equation 5.16 must be positive. Therefore, the objective of placement for reliability is to ensure that interchanging any two modules results in a positive value for Equation 5.16.

Assuming that the hazard rate is a function of temperature, which can be approximated by a Taylor series expansion, Equation 5.16 can be written as:

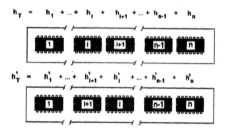

Figure 5.4. Module interchange on a row

$$h'_T - h_T = \sum_{m-1}^{\infty} \{ \sum_{k=1}^{i-1} \frac{dh_i^m(T_i)}{dT^m} \frac{(T'_i - T_i)^m}{m!}$$
$$+ \frac{dh_i^m(T_i)}{dT^m} \frac{(T'_i - T_i)^m}{m!}$$
$$- \frac{dh_{i+1}^m(T'_{i+1})}{dT^m} \frac{(T_{i+1} - T'_{i+1})^m}{m!}$$
$$+ \sum_{k=i+2}^{N} \frac{dh_k^m(T_k)}{dT^m} \frac{(T'_k - T_k)^m}{m!} \} \qquad (5.17)$$

In the following sections, examples of the hazard-rate placement theory are provided for both the two heat-sink and the single heat-sink conduction problems. Equation 5.17 is used as the starting point for the placement methods.

Example: Consider the problem of reliability placement for a conduction-cooled substrate with two opposing heat sinks of equivalent temperatures, as depicted in Figure 5.5.

Figure 5.5. Two-edge conduction placement model

To determine the hazard rate for each module on the substrate, it is necessary to calculate the module temperatures of the original and new placement configurations. This requires the calculation of the associated substrate temperature. The placement of a module on a conductively cooled substrate affects the temperature of all the other placed modules. For a single row, the steady-state temperature at any location can be determined by adding the temperature contributions of all participating factors. To determine the thermal contribution of the mth module at any position, x, in a row of modules (See Figure 5.6), consider the mth module to be a point source dissipating heat at a rate of q_m, and thus make first-order approximations.

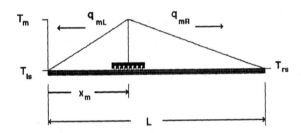

Figure 5.6. Temperature contribution of module m

For the simplest case, assume T_{ls} and T_{rs} are equal. Under these assumptions, the temperature contribution of the mth module to the temperature at any position x is

$$T_m = \frac{q_m(x_{mL} - x_m^2)}{kAL} \times MIN(\frac{x}{x_m}, \frac{L-x}{L-x_m}) \qquad (5.18)$$

where k is the thermal conductivity of the substrate, L is the length of the substrate, A is the cross-sectional area of the substrate, and q_m is the rate of heat dissipated by the mth module. The substrate temperature, $T(x)$, is determined by the contribution of every module on the row,

$$T(x) = T_s + \sum_{m=1}^{N} T_m(x) \qquad (5.19)$$

where T_s is the heat-sink temperature of both the left and right heat sinks.

The module temperature T_i of the ith module is equal to the substrate temperature under the module, plus the temperature increase between the local substrate and the module:

$$T_i(x) = T(x) + q_i Rcb_i \qquad (5.20)$$

where Rcb_i is a constant that specifies the thermal resistance between the case and the substrate for module i. Notice that under the assumption that Rcb_i is constant for any given module at any given position x, the module temperature is equal to the substrate temperature plus a constant term. Thus, the temperature of any module, i, is a function of its position on the substrate, and the only change in the temperature of a module resulting from moving a module is picked up by the change in the substrate temperature function. The difference between the module temperatures of the new and original placement configurations is

$$T_i' - T_i = T(x_i') + T(x_i) \tag{5.21}$$

Using Equation 5.21, the temperature differences for all modules can be determined. Substituting the temperature differences into Equation 5.17, assuming second and higher order terms are negligible with respect to the first-order term, and averaging the equation, as was done in the convection case, in terms of a priority metric, gives

$$
\begin{aligned}
CDPM_i \;=\; & q_i\{\frac{dh_i(T_i)}{dT}[(1 - \frac{x_{i+1} + x_i}{L}) - \sum_{j=1}^{i-1}\frac{q_j}{q_i}\frac{x_j}{L} - \sum_{j=i+2}^{N}\frac{q_j}{q_i}(\frac{x_j}{L} - 1)] \\
& - \sum_{k=1}^{i-1}\frac{dh_k(T_k)}{dT}(\frac{x_k}{L}) \\
& - \sum_{k=i+2}^{N}\frac{dh_k(T_k)}{dT}(\frac{x_k}{L} - 1)\}
\end{aligned}
\tag{5.22}
$$

The detailed of this derivation can be found in reference [OST89]. If the placement is symmetric with respect to heat dissipation and the first derivative of hazard rate with respect to temperature, Equation 5.17 simplifys to:

$$h_T' - h_T \cong \frac{dh_i(T_i)}{dT}q_i - \frac{dh_{i+1}(T_{i+1}')}{dT}q_{i+1} \tag{5.23}$$

Equation 5.23 allows us to develop a placement procedure that is dependent only on the module under consideration. If h_T is the minimum total hazard rate, the left-hand side of Equation 5.23 must be positive for $x_i < [L - (x_{i+1} - x_i)]/2$, referenced from either heat sink. Thus, a placement scheme can be developed using the priority metric, CDP, defined as:

$$CDP_i = q_i\frac{dh_i(T_i)}{dT} \tag{5.24}$$

Only the approximate temperature of module i is needed to calculate the priority number. The ordering of modules is dependent only on the individual module derivative of the hazard rate with respect to temperature and the heat dissipation rate of the module.

A simple but effective placement procedure for the two-heat-sink row is provided below. In this placement procedure, the conduction placement metric, CDP, given by Equation 5.24, defines the selection criterion. The two-opposing-edge heat-sink placement procedure is illustrated in Figure 5.7.

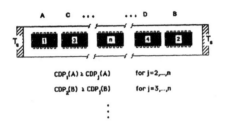

$$CDP_1(A) \geq CDP_j(A) \qquad \text{for } j=2,...,n$$
$$CDP_2(B) \geq CDP_j(B) \qquad \text{for } j=3,...,n$$
$$\vdots$$

Figure 5.7. Two-edge conduction placement rule

Step 1. Assume that the modules can be placed at any position on the row provided overlap does not occur. Initially, all positions are open and all modules are unplaced. Target positions are based on their proximity to the edge heat sinks, and alternate between the two heat sinks. Determine the thermal resistance between the module and the substrate, Rcb, for every module, using a typical resistance-networking technique. The first target location may be adjacent to either the right or left heat sink.

Step 2: Determine the module temperature for each unplaced module at the target position, x_t. The substrate temperature at the target location is approximated by assuming that the heat dissipated by all unplaced modules is treated as coming from a single source located at the center of the substrate. Thus, the module temperature at the target location is

$$T_j(x_t) = \frac{(C_1 - q_j)}{2kA}MIN(x_t, L - x_t)$$
$$+ C_2\frac{(L - x_t)}{kAL} + C_3\frac{x_t}{kAL}$$
$$+ \frac{q_j x_t(L - x_t)}{kAL} + q_j R_{cb_j}$$
$$+ T_s \tag{5.25}$$

where C_1 is the sum of the heat dissipation rates of all unplaced modules, and q_j is the heat dissipation rate of the module under consideration. C_2 is defined as:

$$C_2 = \sum_{k \in A} q_k x_k \qquad (5.26)$$

with A being the set of placed modules on the range $0 \leq x_k \leq L/2$. C_3 defined as

$$C_3 = \sum_{k \in B} q_k(L - x_k) \qquad (5.27)$$

with B defined as the set of placed modules on the range $L/2 \leq x_k \leq L$.

Step 3: Calculate the CDP for every unplaced module, based on the previously evaluated temperatures.

Step 4: Select the module with the maximum CDP and assign the selected module to the target location. In the case of a tie, an arbitrary selection may be made. Update C_1, C_2, and C_3.

Step 5: Select the next target location. Target locations should alternate between the two heat sinks and be based on proximity to the current heat sink. Repeat steps 2 through 5 until all modules are placed.

Example: Consider placement on a substrate with a single constant temperature heat sink based on minimization of the total hazard rate is addressed. The substrate is assumed to be cooled by a single-edge heat sink, with the other three edges insulated, as depicted in Figure 5.8. The analysis begins with the development of thermal equations used to approximate the module temperatures on the substrate.

Figure 5.8. Single-edge conduction placement model

⬦

For a single row, the steady-state temperature at any location on the row is a function of the heat-sink temperature and the thermal contributions of all heat dissipating modules on the row. Consider the thermal contribution of a module m located at any position, x_m, measured from the heat-sink position on the row.

If we consider module m to be a point heat source dissipating heat at a rate of q_m, then the temperature of module m is given by

$$T_m(x_m) = \frac{q_m x_m}{kA} + T_{ls} \tag{5.28}$$

where k is the thermal conductivity of the substrate, A is the cross-sectional area of the substate, and T_{ls} is the edge heat-sink temperature.

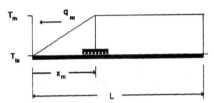

Figure 5.9. Single-edge conduction temperature contribution of module m

The temperature contribution of module m at any position x, as depicted in Figure 5.9, is given by:

$$T_m(x) = q_m R_m MIN(\frac{x}{x_m}, 1) \tag{5.29}$$

where $R_m = x_m/(kA)$. The substrate temperature at any location, x, is given by the heat-sink temperature and the temperature contribution of every module located on the row, as in Equation 5.29. Assuming a constant heat resistance path between the substrate and the module, the module temperature of the kth module, T_k, is given by Equation 5.20. By determining the temperature differences resulting from interchanging two adjacent modules, i and $i + 1$, and substituting into Equation 5.17, assuming the second and higher order terms are negligible with respect to the first order terms gives:

$$\frac{h'_T - h_T}{\Delta R} = \frac{dh_i(T_i)}{dT}(q_i + \sum_{j=i+2}^{N} q_j)$$
$$+ \frac{dh_{i+1}(T'_{i+1})}{dT}(q_{i+1} + \sum_{j=i+2}^{N} q_j)$$
$$+ \sum_{j=i+2}^{N} \frac{dh_j(T_j)}{dT}(q_i - q_{i+1}) \tag{5.30}$$

Restricting Equation 5.30 to a positive value, and ignoring the heat dissipation rates of the modules between x_{i+1} and the insulated edge, a placement metric can be developed of the form:

$$SHP = \frac{dh_i(T_i)}{dT} q_i \qquad (5.31)$$

A simple placement procedure for a single-edge heat sink is outlined in the following section. The single-edge conduction placement procedure is illustrated in Figure 5.10.

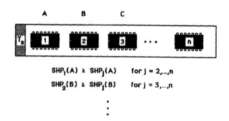

Figure 5.10. Single edge conduction placement rule

◇

Step 1. Assume that the modules can be placed at any position on the row, provided overlap is not acceptable. Initially, all positions are open. Target positions are based on their proximity to the edge heat sinks. Determine the thermal resistance between the module and the substrate, Rcb, for every module, using a typical resistance-networking technique. The first target location is adjacent to the left heat sink.

Step 2: Determine the module temperature for each unplaced module at the target position, x_t. The substrate temperature at the target location is approximated by assuming that the heat dissipated by all unplaced modules is treated as originating from a single source located adjacent to the target location closest to the insulated edge. Thus, the module temperature at the target location is

$$T_i(x_t) = \frac{(C_1 - q_i)x_t}{kA} + \frac{q_i x_t}{kA} + C_2 + q_i Rcb_i + T_s \qquad (5.32)$$

where C_1 is the sum of the heat dissipation rates of all unplaced modules and q_i

is the heat dissipation rate of the module under consideration. C_2 is defined as:

$$C_2 = \sum_{j \in B} \frac{q_j x_j}{kA} \tag{5.33}$$

with B as the set of placed modules in the range $0 \le x_j \le x_t$.

Step 3: Calculate the SHP for every unplaced module, based on the previously evaluated temperatures.

Step 4: Select the module with the maximum SHP and assign the selected module to the target location. If two or more modules have the same SHP, an arbitrary selection may be made. Update C_1 and C_2.

Step 5: Select the next target location. The selection of target locations proceeds from the heat sink edge towards the insulated edge. Repeat steps 2 through 5 until all modules are placed.

5.1.3 Placement on substrates

In the previous sections, placement for reliability was examined for various cooling technologies on a given row. However, the actual placement problem deals with the issue of placement on the entire substrate. In order to place modules on a substrate, the substrate is divided into rows that are parallel the heat-flow direction, and into columns that are perpendicular to the heat-flow. The intersection of the rows and columns forms slots that can accommodate a single module. Initially, the modules are resized to fit in one slot or several adjacent slots, depending on the disparity in the original module sizes. The modules are resolved back to their actual sizes upon completion of placement. The resizing process attempts to maintain actual module sizes to facilitate the resolution process.

As in row placement, target locations are selected based on their proximity to the heat sinks. For convection cooling, the heat sink is represented by the inlet edge of the substrate. The initial target locations are represented by the slots adjacent to the heat sink. To perform the placement procedure, the thermal resistances and the hazard rate equations for each module must be defined. In addition, the thermal properties and characteristics must also be specified.

Initially, all modules are assumed to be unplaced and all slots are open. The thermal analysis is performed on a row-by-row basis. The placement metrics for reliability are defined based on the cooling technology. For conduction-cooled substrate, the placement metric is defined by Equation 5.24. For convection cooling, the placement metric is defined by Equation 5.15. Module temperatures are approximated based on the modifications of the equations provided in the row-placement procedures outlined previously. For convection cooling, Equation 5.5 is used without modifications. For the two-edge heat-sink problem, Equation 5.25 is modified by replacing the row cross-sectional area for the temperature contribution of the unplaced modules with the entire substrate cross sectional area. Similarly, for the single-edge heat-sink problem, Equation 5.32 is modified by replacing the row cross-sectional area by the entire substrate cross-sectional area in the unplaced temperature contribution term. Temperatures are calculated for each module, and the appropriate placement metric is determined for each.

Modules with the highest placement metric are then assigned positions in target locations adjacent to the heat sink. For the two-edge conduction problem, the target locations include both heat sinks. The modules with the highest placement metric alternatively fill both heat-sink edges. The process for equalizing the heat in each row, necessary for all cooling technologies, is accomplished by tabulating the heat dissipation in each row. Selected modules are assigned to rows based on their heat dissipation rates and that of the row. The objective is to appropriately distribute the heat. Once the first column is filled, selection of the next target location depends on the total heat dissipation level of each row. Successive target locations are based on the open slot closest to the inlet on the row with the lowest heat-dissipation level. If a row becomes filled, the row with the next lowest heat-dissipation level is selected. The heat-dissipation level in each row is recalculated after each placement.

5.2   Placement for Fatigue Dependent Reliability

In certain cases, the time to failure may be identified as a critical factor in reliability. For example, with thermal cycling, placement should address the specific mechanism and improve placement based on the critical modules.

The parameter used in examining failures due to fatigue is the mean number of cycles to failure, N_f, often modeled by a power law equation:

$$N_f = A(K\Delta T)^\alpha \tag{5.34}$$

where A, K, and α are based on environmental, geometric, and material properties. The value of α represents the slope on a log-log plot of N versus $(K\Delta T)$ and has a negative value. A specific example of this equation for solder joint fatigue is provided below.

Example: Consider the case of four J-leaded modules named U_1, U_2, U_3 and U_4, with power dissipation rates 1.0, 0.8, 0.6 and 0.4 (watts), respectively. The modules are to be placed on a convection-cooled row. The substrate material is assumed to be alumina, and the modules are plastic-packaged devices with the same geometries. The solder joint life of the modules is based on a thermal coefficient of expansion mismatch between the module case and substrate; temperature and/or power cycling therefore produces cyclic inelastic strains in the solder joint. The magnitude of the strain amplitude can be computed based on the mismatch, and the magnitude of the stress amplitude can be estimated from the spring stiffness of the module lead configuration. Subsequent creep and/or stress relaxation leads to a hysteresis loop, which gives a measure of the total inelastic energy dissipated in the solder. The amount of stress relaxation depends on the frequency of loading, dwell times, stress, and temperature history throughout the cycle and during dwells. The fatigue life can then be related to the energy dissipated, using an approximate form of the Coffin-Manson law, provided the fatigue properties of the solder material are well characterized. The effects of temperature and dwell times are included by suitable modifications to the fatigue properties of the material. The equation below is utilized for fatigue life estimation, N_f, for plastic J-lead packages on alumina substrate.

$$N_f = \frac{1}{2}\left[\frac{K(L_D\Delta\alpha\Delta T_e)^2}{400\epsilon'_f\ A\ h}\right]^{\frac{1}{c}}$$

where
- $K = 240$
  (empirical factor relating energy dissipated in hysteresis and spring stiffness of the J-lead);
- $A = 0.001\ inch^2$
  (footprint area of for solder joint);
- $h = 0.015\ inches$
  (Solder joint height);

- $\Delta\alpha = \alpha_{alumina} - \alpha_{plastic} = (6 - 90)/10^6 = -84 \times 10^{-6}$;
- $\epsilon'_f = 0.325$

  (material fatigue constant for 60 - 40 Sn-Pb solder);
- $C = -0.442 - 1.5 \times 10^{-4}(2T_{c_i} + 2T_o) + 1.72^{-2}ln(1 + 360/dwell\ time)$

  (material fatigue parameter);
- T_{s_i} = substrate temperature;
- T_{c_i} = case temperature of the ith module;
- T_0 = base temperature;
- $\Delta T_e = T_c - T_o$ by assuming $T_{s_i} = T_{c_i} - T_c$.

Assuming the module case and substrate temperatures are approximately the same,

$$T_{C_i}(S) = T_f(S) + R_{cf_i}q_i$$

where $R_{cf_i} = 4°C/\text{Watt}$. The row is cooled by air with an inlet temperature of $50°C$ at a rate of 0.002 kg/sec. The row is divided into four slots into which any module can be placed. the position closest to the inlet is designated A, followed by B, C, and D. According to the placement theory, we determine the fluid temperature and module at position D: $T_f(D) = (1.0+0.8+0.6+0.4)/(1,000\times 0.002) + 50 = 51.4°C$, $T_{c_{U_1}} = 55.4°C$, $T_{c_{U_2}} = 54.6°C$, $T_{c_{U_3}} = 53.8°C$, and $T_{c_{U_4}} = 53.0°C$. Based on the fatigue equation and the module temperatures, module U_4 has the highest life expectancy. Thus, U_4 is selected and positioned at D. The module temperatures at position C are then evaluated, and the fatigue life of modules U_1, U_2, and U_3 are determined. In this case, module U_3 has the maximum life and is positioned at C. At the B position, module U_2 is selected, and module U_1 is positioned at A.

◇

In fatigue analysis, reliability is reported in terms of the variability of the number of cycles to failure, as given by Equation 5.34. For systems in which a device failure results in an entire failure, the weakest modules become the most critical. Placement of modules based on fatigue failures requires that the modules with minimum mean cycles to failure are successively maximized. Mathematically stated, the placement problem becomes

$$MAX(MIN(N_{f_i})) \quad for\ i = 1, ..., n \qquad (5.35)$$

where n is the number of modules to be placed.

For convection cooling, the module with the smallest number of mean cycles to failure should be placed at inlet edge of the substrate. However, the effect of positioning modules upstream is felt by all downstream modules. Thus, the selection of module and position on a row must consider the heat dissipation rate of the module as well as the mean cycles to failure. For example, if the coefficients of N_f and the thermal resistance path to the working fluid are identical for all modules on a row, the modules should be placed in ascending order of their heat dissipation rates.

For the convection-cooling model, the placement of modules on a row can be performed by examining the worst position. Because the mean cycles to failure equation is a decreasing function of the cyclic temperature difference, the worst position is at the outlet edge of the row. The fluid temperature at the end of the row can be approximated by:

$$T f_{outlet} = T f_{inlet} + \frac{1}{\dot{m} C_p} \sum_{j=1}^{N} q_j \qquad (5.36)$$

The technique examines target locations based on their proximity to the outlet edge of the substrate. As in hazard-rate placement, the thermal resistance between the module and the working fluid must be determined. In addition, the parameters of the number of mean cycles to failure equation must be known.

Once the required information has been established, the cyclic temperature difference is calculated for each module at the outlet position, using the fluid temperature defined in Equation 5.36. In addition, the number of mean cycles to failure for each module is approximated by the estimated cyclic temperature difference. The module with the highest number of mean cycles to failure is selected and placed adjacent to the outlet edge of the substrate. If two modules have the same maximum number of mean cycles to failure, the module with the highest heat dissipation rate is selected. The heat dissipation of the selected module is removed from the sum in Equation 5.36 and the new fluid temperature is determined. The position immediately upstream of the previously placed module is selected. Individual cyclic temperature differences are then reevaluated, and the number of mean cycles to failure is determined for each unplaced module. Once again the module with the highest number of mean cycles to failure is selected, and the heat dissipation rate of the selected module is removed from the sum in Equation 5.36. The process continues until all modules are placed on the row (See

Figure 5.11).

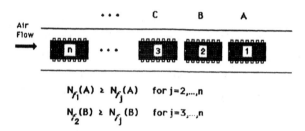

Figure 5.11. Time-to-failure placement rule for forced -convection case

For an entire substrate, the placement process is similar to that of an individual row. In this case, the total heat dissipated by all the modules to be placed on the substrate represents the sum of the heat dissipation rates in Equation 5.36; the mass flow rate is for the entire substrate. Cyclic temperature differences are then calculated for every module, based on the average outlet temperature. Using the calculated cyclic temperature differences, the number of mean cycles to failure for each module is determined. Modules with the highest number of mean cycles to failure are selected and placed along the outlet edge of the substrate. The heat dissipation in each row is recorded, and the heat dissipation rates corresponding to the selected modules are removed from the sum of heat dissipation rates given by Equation 5.36.

After the heat dissipation rates of the placed modules are removed, the new average fluid temperature is determined for the column immediately upstream of the previously place modules. The new cyclic temperature difference and the number of mean cycles to failure is then calculated for the remaining unplaced modules. Once again, the modules with the highest number of mean cycles to failure are selected. The selected modules are placed in rows based on their heat dissipation rates. From the selected set of modules, the module with the highest heat dissipation rate is placed in the row with the lowest heat dissipation rate. In this manner, an even distribution of heat can be maintained across the substrate. Once the column adjacent to the initially placed modules has been filled, the heat dissipated by the placed modules is removed from the total heat dissipation rate in Equation 5.36. The process continues from the outlet end to the inlet end until all modules are placed on the substrate.

5.3   Placement for Vibration[1]

Transportation, handling, and the application environment often induce vibrations in the substrate assembly. Severe or long-term vibration environments can cause fracture and fatigue failures of electronic systems. Common failures include broken wires, fractured substrates, broken module leads and cracked solder joints [STE88].

Failure caused by vibration can be decreased by reducing the vibration-causing deflection amplitudes and stresses. Several commonly used strategies include adding stiffening ribs, reducing the overall size of the substrate, making the boundary conditions more rigid, and rearranging the modules. The introduction of additional ribs usually increases weight and cost, restricts area, and may not be feasible when electronic routing is considered. Because the geometry of the substrate is usually minimal to begin with or is fixed due to interchangeability requirements, it is generally impossible to reduce the substrate size. Changing the boundary conditions of the substrate by gripping more securely can be a reasonable strategy as long as manufacturing costs stay within bounds, but there are limitations in terms of the benefits for large and/or thin substrates, where interior deformation may still be large.

Another approach to reducing vibrationally induced failures is to appropriately locate modules to increase the fundamental natural frequency of the substrate assembly. In order to do this, the design team needs a method to calculate the natural frequencies and mode shapes of the substrate assembly for all combinations of boundary conditions, a computational-efficient objective function that can provide a basis for choice between alternative and original placements, and an optimization method to determine whether the objective function has increased or decreased during the rearrangement process.

5.3.1   Deformation, stress and vibration

While there are many methods to determine the deformation and stress response of substrate assemblies, the finite element method (FEM) is the most commonly used tool. Engel and Lim [ENG86, ENG88] used FEM to perform the static stress analysis of module-lead-substrate (MLS) groups by modeling individual MLSs

---

[1]This section was written with the assistance of Dr. E. Magrab, based on the work of Dr. T. S. Chang [CHA92].

as orthotropic thin-plate elements, the leads and pins as beam elements, and the substrate as orthotropic or isotropic plate elements. Later, Engel [ENG90] proposed an approximate engineering flexural analysis to simplify matters. He modeled the MLSs as strips in both the x- and y-directions in order to eliminate the cumbersome experimental work required to determine the material constants on the microstructure scale. Wong [WON90] combined an analytical analysis with experimental load-deformation characteristics of assemblies to predict the maximum allowable loadings and deflections that the mounted modules could withstand before incurring failures. Engel [ENG91] also experimentally measured the effective stiffness of a module-reinforced circuit card to show the effect on reliability. Pitarresi [PIT91] applied a smeared-out process originally presented by Cheng [CHE82], Olhoff [OLH81] and Bendsøe [BEN81] to model the circuit card assemblies and to determine its equivalent material constants experimentally. Chang and Magrab [CHA91] also presented a procedure for the determination of the elastic constants of assemblies.

In using finite-element methods to determine the natural frequencies for substrate assemblies, computing all of the system's natural frequencies and mode shapes is often expensive and time-consuming. Several methods have been introduced to minimize the computational time, while still retaining enough of the system's characteristics to accurately obtain the lowest natural frequencies [IRO65, GUY65, KID73, KID75, PAZ83, PAZ84]. These methods result in so-called reduced systems. Depending on their assumptions, these methods can be classified as either static or dynamic condensation methods.

The first reduction method, presented by Guyan in 1965 [GUY65], is based on the static condensation of unwanted and/or external force-free coordinates. Guyan's method can determine static displacements accurately. When applying Guyan's method to a dynamic system, the accuracy of the natural frequencies and mode shapes depends on how appropriate the chosen reduced system — defined by the relation of master degrees of freedom (user-selected coordinates) — is to the original system. If the degrees of freedom are not properly chosen, large errors can result, especially when the associated inertial forces are large. Guyan's method also requires a considerable number of inverse matrix operations.

A modified reduction method, presented by Kidder [KID73, KID75], improves the accuracy of the mode shapes by retaining more information when formulating the reduced system. Nevertheless, the accuracy of the natural frequencies still depends on how well the degrees of freedom are chosen. Another dynamic condensation method was introduced by Paz [PAZ83, PAZ84]. Paz's method can be used to accurately determine the natural frequencies and mode shapes using only a few iterations.

For the substrate assembly, the magnitude of the transverse deflections is generally much larger than that of the rotational deflections. By choosing all the transverse deflection degrees of freedom of the substrate assembly model as the master degrees of freedom, the fundamental natural frequency can be determined from a reduced system that is approximately one-third the size of the original problem.

5.3.2 Modeling for automatic rearrangement

In placement for vibration, the individual modules are generally replaced by equivalent orthotropic plates, called the module-lead-substrate assembly (MLS). The challenge is to characterize these MLSs in such a manner that they adequately represent reality, while still providing a reasonable means for their automatic rearrangement during the optimization process.

In order for the rearrangement of the MLSs to be feasible, the following boundary constraints are required:

$$a_i/2 \leq x_i \leq a - a_i/2$$
$$b_i/2 \leq y_i \leq b - b_i/2$$

$$(5.37)$$

where $i = 1, ..., N_c$, N_c is the total number of MLSs comprising the assembly, a is the largest length of the substrate along the x-direction, b is the largest length of the substrate along the y-direction, x_i, y_i are the coordinates of the center of the ith MLS, a_i is the length of the ith MLS, and b_i is the width of the ith MLS (see Figure 5.12). To prevent overlapping of a rearranged (swapped) MLS with another MLS, one of the following equations must also be satisfied:

$$|x_i - x_j| \geq (a_i + a_j)/2$$
$$|y_i - y_j| \geq (b_i + b_j)/2$$

$$(5.38)$$

where $i = 1, ..., N_c$. To simplify the feasibility checking process, the placement problem is transformed into a finite element model that is used to develop an automatic rearrangement algorithm.

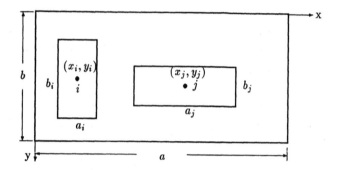

Figure 5.12. Notation of the MLS placement approach

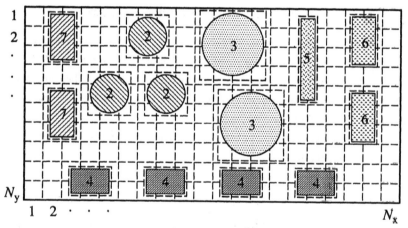

Figure 5.13. Mesh for the substrate assembly

As an example consider the assembly shown in Figure 5.13. The numbers 2 through 7 represent different types of modules mounted on the substrate. Each individual module of the assembly is modeled as either a rectangular or a square MLS, as shown in Figure 5.13. The total number of MLSs to be rearranged is N_c. The approximate dimensions $(a \times b)$ of the substrate and those of the individual MLSs $(a_i \times b_i,\ i = 1, ..., N_c)$ are used to determine the size of the mesh for the finite element model. Once the size of the mesh is determined, the substrate is overlaid with a grid of square elements, as shown in Figure 5.13. The introduction of this mesh is designed to reduce the complexity of the automatic rearrangement process, while still retaining a reasonable means of describing the geometry of the individual MLSs.

To avoid the possibility of any of the MLSs overlapping in the final configuration obtained from the automatic rearrangement process, the size of each individual MLS is enlarged to conform to its nearest exterior grid line, as shown in Figure 5.14. Those MLSs that have dimensions equal to an integer multiple of the size of grid k are relocated to the nearest grid line (see Figure 5.14).

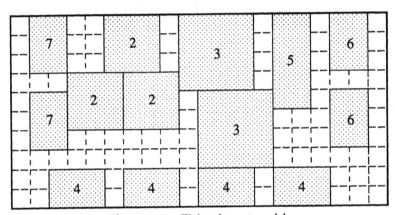

Figure 5.14. Finite element model

Two arrays, $\{P\}$ and $\{S\}$, give the lengths, l_i, and widths, w_i, respectively, of the corresponding MLS type i. For this example, $\{P\} = \{l_1, l_2, ..., l_{N_d}\} = \{1, 3, 4, 3, 2, 2, 2\}$. A two-dimensional array, called the content matrix C, is constructed to store the corresponding type number of each individual element of the current configuration. Thus $C_{ij} = B_n$, where $n = 1, 2, ..., N_d$. For the assembly shown in Figure 5.14, C is given by:

$$
C = \begin{bmatrix}
1 & 7 & 7 & 1 & 1 & 2 & 2 & 2 & 1 & 3 & 3 & 3 & 3 & 1 & 5 & 5 & 1 & 6 & 6 & 1 \\
1 & 7 & 7 & 1 & 1 & 2 & 2 & 2 & 1 & 3 & 3 & 3 & 3 & 1 & 5 & 5 & 1 & 6 & 6 & 1 \\
1 & 7 & 7 & 1 & 1 & 2 & 2 & 2 & 1 & 3 & 3 & 3 & 3 & 1 & 5 & 5 & 1 & 6 & 6 & 1 \\
1 & 1 & 1 & 2 & 2 & 2 & 2 & 2 & 2 & 3 & 3 & 3 & 3 & 1 & 5 & 5 & 1 & 1 & 1 & 1 \\
1 & 7 & 7 & 2 & 2 & 2 & 2 & 2 & 2 & 1 & 3 & 3 & 3 & 3 & 5 & 5 & 1 & 6 & 6 & 1 \\
1 & 7 & 7 & 2 & 2 & 2 & 2 & 2 & 2 & 1 & 3 & 3 & 3 & 3 & 1 & 1 & 1 & 6 & 6 & 1 \\
1 & 7 & 7 & 1 & 1 & 1 & 1 & 1 & 1 & 1 & 3 & 3 & 3 & 3 & 1 & 1 & 1 & 6 & 6 & 1 \\
1 & 1 & 1 & 1 & 1 & 1 & 1 & 1 & 1 & 1 & 3 & 3 & 3 & 3 & 1 & 1 & 1 & 1 & 1 & 1 \\
1 & 1 & 4 & 4 & 4 & 1 & 4 & 4 & 4 & 1 & 4 & 4 & 4 & 1 & 4 & 4 & 4 & 1 & 1 & 1 \\
1 & 1 & 4 & 4 & 4 & 1 & 4 & 4 & 4 & 1 & 4 & 4 & 4 & 1 & 4 & 4 & 4 & 1 & 1 & 1
\end{bmatrix}
$$

A one-dimensional location array $\{L\}$ whose length is equal to the total number of MLSs, N_x, records the upper-left-corner address number in the content matrix C, corresponding to each individual MLS type for the current configuration. For example, Figure 5.15 shows that $\{L\} = \{1, 2, 4, 5, 6, 9, 10, 14, 15, 17, 18, 20, 21, 24, 25, ..., 194, 198, 199, 200\}$. Thus, the grid number of the upper-left-corner of the individual MLSs for the current placement is registered as a single integer number. Each number corresponds to an MLS type, determined from the content matrix C, and dimensions from the $\{P\}$ and $\{S\}$ arrays. In this manner, the connectivity is maintained during the placement process. For each L_k selected, the coordinates of the corresponding element in C are

$$i = 1 + INTEGER[(L_k - 1)/N_x]$$
$$j = L_k - (i-1)N_x \qquad (5.39)$$

1	2		4	5	6			9	10				14	15		17	18		20
21			24	25				29					34			37			40
41			44	45				49					54			57			60
61	62	63	64			67							74			77	78	79	80
81	82								90	91						97	98		100
101									110					115	116	117			120
121			124	125	126	127	128	129	130					135	136	137			140
141	142	143	144	145	146	147	148	149	150					155	156	157	158	159	160
161	162	163			166	167			170	171			174	175			178	179	180
181	182				186				190				194				198	199	200

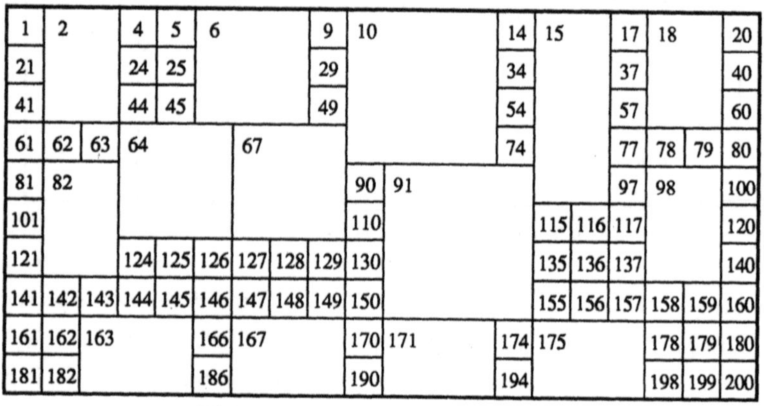

Figure 5.15. Location index array

The assembly has now been transformed into a finite element model consisting of square orthotropic elements, and its configuration can be fully described by the associated content matrix C, the location index array $\{L\}$ and the equivalent MLS dimensions in array $\{P\}$ and $\{S\}$. The introduction of the location index array reduces the size of the rearrangement process and can handle the special situations where certain MLSs have been designated immovable or where certain areas are prohibited from accepting swapped MLSs. The requirements imposed by the feasibility of the size of the MLS and the boundary and containable constraints assure that the interchange process will result in a feasible configuration of the assembly for any combination of equal, unequal, movable and immovable MLSs.

5.3.3 Placement algorithm

A general algorithm by which the above placement process can be performed automatically follows. For each MLS interchange attempt, the feasibility of the size of the MLS and the boundary, non-overlapping, and containable constraints are checked. A successful swap is obtained when all the constraints are satisfied and the fundamental natural frequency has been increased. The flowchart for the automatic placement process of the assembly is given in **Figure 5.16**. Although not explicitly shown in this figure, the algorithm is based on the simulated annealing method.

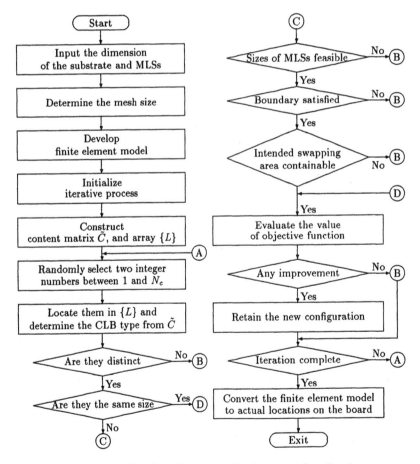

Figure 5.16. Flowchart for automatic placement for vibration

To determine whether or not the fundamental natural frequency has been increased after interchanging the MLSs, the objective function, F', is used:

$$- F' = \omega_1'^2 = \frac{U_0 + (U_1^2 - U_1^1) + (U_2^2 - U_2^1)}{\hat{T}_0 + (\hat{T}_1^2 - \hat{T}_1^1) + (\hat{T}_2^2 - \hat{T}_2^1)} \qquad (5.40)$$

where

$$U_0 = \sum_{i=1}^{N_y} \sum_{j=1}^{N_x} U_{ij}(C_{ij})$$

$$\hat{T}_0 = \sum_{i=1}^{N_y} \sum_{j=1}^{N_x} \hat{T}_{ij}(C_{ij})$$

$$U_s^t = \sum_{i=I_s}^{I_s+\omega-1} \sum_{j=J_s}^{J_s+l-1} U_{ij}(C_{ij}^t) \qquad (5.41)$$

$$\hat{T}_s^t = \sum_{i=I_s}^{I_s+\omega-1} \sum_{j=J_s}^{J_s+l-1} \hat{T}_{ij}(C_{ij}^t) \qquad (5.42)$$

$$U_{ij}(C_{ij}) = (D_x)_{C_{ij}}(S_1)_{ij} + (D_c)_{C_{ij}}(S_2)_{ij} + (D_y)_{C_{ij}}(S_3)_{ij} + (D_{xy})_{C_{ij}}(S_4)_{ij}$$

$$\hat{T}_{ij}(C_{ij}) = (\rho h)_{C_{ij}}(S_5)_{ij}$$

$$(S_1)_{ij} = (\frac{\partial^2 W_1}{\partial x^2})_{ij}^2$$

$$(S_2)_{ij} = 2(\frac{\partial^2 W_1}{\partial x^2})_{ij}(\frac{\partial^2 W_1}{\partial y^2})_{ij}$$

$$(S_3)_{ij} = (\frac{\partial^2 W_1}{\partial y^2})_{ij}^2$$

$$(S_4)_{ij} = 4(\frac{\partial^2 W_1}{\partial x \partial y})_{ij}^2$$

$$(S_5)_{ij} = (W_1^2)_{ij}$$

The quantity, $W_1 = W_1(x, y)$, is the original plate's mode shape corresponding to its fundamental natural frequency ω_1, ρ and h are the mass density per unit volume and thickness of the plate, respectively, and $D_x, D_y, D_{x,y}$ and D_c are the flexural rigidities of the orthotropic plate. The subscripts i and j indicate the location of each $k \times k$ element in the $\hat{C}$ matrix. The subscript C_{ij} denotes the material properties corresponding to that element. C_{ij}^t, $t = 1, 2$ denotes the element matrix $\hat{C}$ for the current and new configurations, respectively, and $s = 1, 2$ denotes the location indices for the two sets of elements to be exchanged.

For moderate changes in the properties of an MLS structure, it is reasonable to assume that $W_1(x, y)$ does not change. Furthermore, the second derivatives of the transverse displacement can be approximated by the appropriate linear combination of the modal displacements determined from the finite difference method. These linear combinations are a function of the plate's boundary conditions for elements on the plate's perimeter. If W_1 is assumed to remain constant throughout the design changes, these approximations to the second derivatives are only computed once. Therefore, $(S_k)_{ij}$, $k = 1, 2, ..., 5$, is constant.

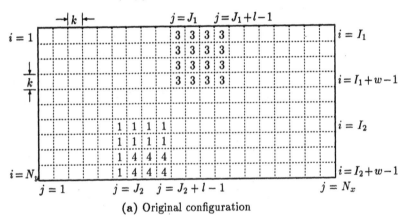

(a) Original configuration

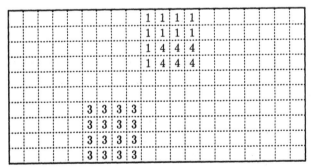

(b) Swapped configuration

Figure 5.17. An example of an interchange between two areas

U_0 and $\hat{T}_0$ are computed only one time. All subsequent computations involve only U_s^t and $\hat{T}_s^t$. In Equations 5.50 and 5.51, two sets of $l \times w$ elements are assumed to be exchanged. One set has the original location of its elements defined by the indices $i = I_1, ..., I_1 + \omega - 1$ and $j = J_1, ..., J_1 + l - 1$, whereas the second set is defined by $i = I_2, ..., I_2 + \omega - 1$ and $j = J_2, ..., J_2 + l - 1$ (see Figure 5.17).

The objective of the design modification is to rearrange these modules in such a way that the fundamental natural frequency of the assembly of modules on a substrate is increased. Therefore, if the objective function, F, of the original configuration is $F = -U_0/\hat{T}_0$, a comparison can be made between a pair of configurations by simply computing the difference in the objective functions.

Example: Suppose that during the placement process, two sets of elements to be interchanged, are the two $l \times \omega$ areas shown in Figure 5.17. The justification for accepting the interchange of these two areas requires only the computation of the quantities shown in Equations 5.50 and 5.51 with $l = 4, \omega = 4, J_1 = 10, I_1 = 1, J_2 = 6$, and $I_2 = 7$. The value of the objective function for the new configuration, F', is determined from Equation 5.40; that is, $F' = -U_a/\hat{T}_a = -\omega_1'^2$. The fundamental natural frequency for the swapped configuration has increased when $F' < F$. In this case, the swapped configuration is retained and the values of U_0 and $\hat{T}_0$ are set equal to U_a and $\hat{T}_a$, respectively. The current configuration is restored when $F' \geq F$ and the quantities U_0 and $\hat{T}_0$ remain the same..

◇

5.4   Placement for Producibility

The key to placement for producibility is understanding equipment capabilities and manufacturing processes up-front in the design process. By identifying and planning for the abilities of the equipment, as well as relevant formats, configuration options, and process steps, designs will be able to incorporate characteristic of their device mixes and the number of unique formats will subsequently decrease. In this way, process repeatability is improved and production throughput increases through reduced downtime for equipment re-formatting.

Figure 5.18, a typical process flow diagram, shows the various stages from design to manufacture of an assembly. The circuit schematic is provided by the electrical engineering team, and represents the devices and the flow of signals that ensure the functionality of the circuit. The initial selection of devices is focused primarily on satisfying the functional requirements of the electronic circuit design.

The physical structure of the module is determined by available assembly techniques. The device locations and orientations are then determined, and finally the layout is completed by routing. The initial layout is also analyzed by simulation tools that perform thermal, vibrational and reliability analyses, and modifications are made to the initial layout. After the assembly design is completed, the manufacturing process commences, modules are assembled on the substrate, and the assembly is tested.

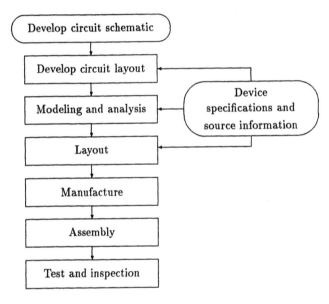

Figure 5.18. Simplified flow diagram of the assembly development process

Tables 5.1 through 5.3 show some of the database information required to establish the link between design and manufacture. Table 5.1 presents the *electronic circuit assembly design criteria*, which is a collection of information that defines the configuration and operation of the completed assembly. Table 5.2 presents the *module specifications and sourcing information*, which is used by both the design and manufacturing team. The design team uses these data to select parts and a layout that best suit the production processes. The manufacturer uses the information about module style and dimensions for product assembly. Table 5.3 presents *placement specifications and information*, which contains information required to assemble the electronic circuit assembly (ECA).

Table 5.1. Design criteria

- Design function specifications:
  - Design/assembly restrictions
  - Module types
  - Substrate configuration
  - Manufacturing processes
  - Module reliability
  - Product reliability
- Temperature power requirements
- Vibration requirements
- Currents and voltages at test points
- Margin of error analysis
- Worst-case analysis
- Test validations

Table 5.2. Module specifications and sourcing information

- Inventory part number
- Manufacturer part number
- Manufacturer identification
- Part function
- Part operating considerations
  - Operation range
  - Storage range
- Part package type, composition, and dimension
- Number of leads/pins, dimensions, and compositions
- Origin (pin 1, center, other)
- Tool number to handle part
- Feeder type

Table 5.3. Placement assembly information

- Logical placement order/process plan
- Tool number
- Location of tool
- Location of feeder
- Feeder rotation from assembly axes
- Lead forming and cutting criteria (non pre-formed)
- Lead form acceptability criteria (table-mounted vision system)
- Placement coordinates
- Rotation/orientation
- Tool point offsets
- Placement offsets
- Vision landmarks
- Tactile search pattern (for insertion)
- Placement pressure

The layout procedure (Figure 5.19) starts with the generation of the parts list from the design criteria. The selected parts are stored with the relevant information in the part specifications and sourcing information. The aim of the layout procedure is to ensure the least number of manufacturing processes and to restrict the design to a single-sided placement. The decision for a two-sided placement is determined by contrasting the area for modules and the available substrate area. The final area is decided by comparing the areas for footprint, gripper release, and inspection. Once the type of substrate is chosen, the corresponding assembly plans are determined.

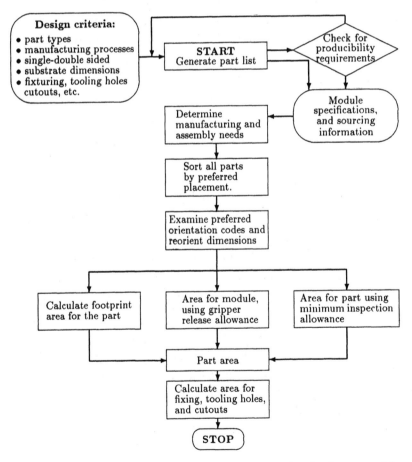

Figure 5.19. Flowchart for generating producible mixed-technology assemblies

Metrics, such as weighted wire length, cycles to failure, and reliability, have been utilized in industry for many years for measuring the success of electronics design under routability. Producibility, on the other hand, generally has no metric assigned to it because it is theoretically possible to assemble any design passed down to the shop floor. The measure of success, then, is yield, cost and schedule. The producibility challenge occurs to use the appropriate equipment to generate high yields in a timely manner.

The costs associated with electronics packaging and production can be classified into three basic categories: overhead, production time, and waste. Overhead is a measure of the costs associated with maintaining the production facility, and conventionally consists of costs such as equipment purchasing, leasing, and repair costs; utilities; facility lease and maintenance; paper work; and advertising.

Production time is a measure of the efficiency of a design in terms of its ability to be manufactured with available production facilities. A design configured with inefficient processes, such as manual soldering, may cost significantly more to produce in large lot sizes than similar designs configured for more efficient *en masse* soldering processes. Additionally, due to the poor process repeatability of manual soldering methods, production runs remain as work in progress for longer periods of time as they are cycled between rework, inspection, and testing stations. Hence, projected increases in production time, which may be directly tied to selected processes and equipment, can be defined as a producibility metric, as it illustrates the direct cost impact of certain design trade-offs.

Waste is a measure of what was bought as raw material for production and was ultimately discarded. Examples of such materials are modules with improperly formed or broken leads, thermally shocked modules, burned and blistered substrates, and substrates with lifted lands. As each of these examples stems from assembly-related problems, or the inability to test the modules until all the modules are placed on the substrate, the percent yield also serves as a producibility metric. Hence, a design that might require a great deal of rework has a high likelihood of generating additional flaws during assembly. If the assembly is scrapped, the waste cost associated with the design increases and the percent yield associated with the design decreases. Hence, more raw materials are required in order to complete the production run.

Both the projected production time and percent yield can be used as a producibility metric. In this way, the module placement equipment and module location on the substrate can be incorporated as part of the design process. If a placement is not feasible under a specific design configuration, the trade-offs in the metrics serve to illustrate which module movements would impact the production yield the least.

5.5   Combined Placement

No single placement can satisfy the large number of goals and constraints that can be applied. In fact, placement for one goal can have an adverse effect on some other goal. Pecht et al. [PEC87] observed this problem for module placement for reliability and routability of printed wiring boards. Any placement procedure must allow for a certain amount of sacrifice on some or all of the placement goals.

In the following sections, combined heuristic placement methods are discussed. As an example, a placement method based on an iterative technique to improve routability and reliability is presented.

5.5.1   Interchange placement

The interchange technique is perhaps the simplest method for combining placement objective. In this procedure, a module is selected as the primary module. This module is interchanged with every movable module on the substrate and the results of the interchange are scored. Scoring is generally based on the amount of improvement in one or more measures of success. If the improvement is significant, the interchange is accepted; otherwise, the modules are returned to their original positions. The interchange method continues until goals are reached, all modules have been interchanged, or some iterative number has been met.

5.5.2   Simulated annealing

The simulated annealing placement procedure is similar to the interchange technique. Given an initial placement, random interchanges are attempted. The total number of interchanges, the number of successful interchanges, and the number of unsuccessful interchanges are tabulated. In addition, the average score and its standard deviation are also calculated. The interchange process continues until the placement configuration reaches equilibrium. Equilibrium is assumed if the total number of interchanges or the number of successful interchanges has reached a predefined limit. Predefined limits are based on the problem size. The random use of the exchange is dependent on an annealing schedule, and is reduced with time. During the entire process, the best configuration is always saved. Theoretically, the best configuration should correspond to the final placement configuration obtained from the annealing process. However, this is often not the case.

The score for reliability placement is often ΔH, or the changed hazard rate. The score for routability placement is often ΔW, or the change in wire length. In this method, module pairs are selected and interchanged, and the results of the interchange are scored. If $\Delta W \leq 0$ and $\Delta H \leq 0$, the interchange is accepted. However, if $\Delta W > 0$ or $\Delta H > 0$, a probability of acceptance is defined. The probability of accepting an increase in wire length is

$$P(\Delta W) = e^{-\omega \frac{\Delta W}{T_W}} \qquad (5.43)$$

where T_W is the parameter for routability placement and ω is the routability weighting factor. The probability of accepting an increase in hazard rate is

$$P(\Delta H) = e^{-(1-\omega)\frac{\Delta H}{T_H}} \qquad (5.44)$$

where T_H is the parameter for reliability placement.

Initially, the starting values of T_W and T_H are experientially determined to provide a high probability of acceptance. This process is analogous to melting the placement configuration. Then, the values of T_W and T_H are reduced according to the annealing schedule.

5.5.3   Force-directed placement

Force-directed placement procedures can be applied to the problem of coupled placement by assuming or determining a placement configuration. A fictitious connectivity matrix is then generated to yield the same placement configuration under force-directed placement processing. The fictitious connectivity matrix is called the position-adjacent matrix, k'_{ij}, because it is derived from an established placement configuration. The position-adjacent matrix is used as the starting point for force-directed placement for various placement attributes, such as reliability. The force equations based on reliability, are superimposed. The method for deriving the position-adjacent matrix is based on examining an existing placement configuration.

Initially, the entire position-adjacent matrix is set to zero and each module is examined individually for its spatial relationship to all other modules. The value of an entry in the position-adjacent matrix, k_{ij}, is determined by the average of the non-zero entries of the connectivity matrix. For example, if a module, i, is adjacent to a module, j, in the x or y direction, then the entry at k'_{ij} is set to ξ. If module i is not adjacent to module j, then the entry at k'_{ii} is set to zero. A repulsive matrix, R'_{ij}, for the position-adjacent matrix can also be established at this time.

The position-adjacent matrix and its associated repulsive matrix are employed to solve for the placement for some placement attributes, much as the connectivity matrix, k_{ij}, and its repulsive matrix, R_{ij}, are used for wire interconnection. For the case of thermal placement, the external forces act on modules in the direction parallel to the primary heat flow. The external force is calculated by

$$Fx_i = -k'_{ii}\Delta x_{is} = -k'_{ii}(x_i - x_s) \tag{5.45}$$

where for the thermal placement case, x_s is the position next to the heat sink or inlet edge for conduction and convection cases, respectively. Δx_{ii} is set equal to Δx_{is} in the placement procedure. The external force is employed to properly orient the placement configuration on the board, regardless of the initial module assignments.

Once the position-adjacent matrix and the repulsive matrix for the various placement attributes are established, a global placement procedure can be performed to establish relative positioning of the modules. This is done through the force-directed placement procedure, using the connectivity matrix and the position-adjacent matrix to calculate the forces on each individual module. Placement is controlled by a weighting factor, ω, which is restricted to the range $0 \leq \omega \leq 1$. The force equations for any movable modules are given by:

$$Fx_i = \sum_{j=1}^{N}[-(\omega k_{ij} + (1-\omega)k'_{ij})\Delta x_{ij} + (\omega R_{ij} + (1-\omega)R'_{ij})\frac{\Delta x_{ij}}{\Delta S_{ij}}] \tag{5.46}$$

and

$$Fy_i = \sum_{j=1}^{N}[-(\omega k_{ij} + (1-\omega)k'_{ij})\Delta y_{ij} + (\omega R_{ij} + (1-\omega)R'_{ij})\frac{\Delta y_{ij}}{\Delta S_{ij}}] \tag{5.47}$$

for $i = 1, 2, ..., M$, where M is the total number of movable modules. The objective in the force-directed approach is to minimize the total force on the modules, such that

$$F_T = \sum_{i=1}^{M}(|Fx_i| + |Fy_i|) = 0 \qquad (5.48)$$

This set of coupled equations is solved by a modified Newton-Raphson method for nonlinear equations. In a sense, the two superimposed placement attributes are allowed to compete with each other for dominance in the placement configuration. The weighting factor allows the designer to bias the final placement configuration outcome for routability (i.e. $\omega = 1$) or for reliability (i.e., $\omega = 0$) based on the desired application.

One problem with the force-directed procedure is that the total force on each module is rarely equal to zero. Therefore, determining an equilibrium position is often difficult. To alleviate this problem, placement is performed by manipulating the modules on the established grid system to eliminate overlap. User intervention techniques are outlined by Osterman [OST90]. In addition, placement on the grid system eliminates the need for the repulsive force that was proposed to facilitate separation between modules. The placement algorithm can also be constrained to operate on the vertical and horizontal forces separately. Thus, movement can be restricted in the x or y directions.

Another problem with the force-directed method, pointed out by Osterman [OST90], is a loss of physical reality during the combined force directed placement stage. Because the force-directed placement is based on an established placement configuration, the placement measure may not be adequately represented. In order to resolve this problem, placement metrics should be employed. For example, in thermal placement, external forces could be directed towards the heat sink(s) and be proportional to the modules' thermal placement metric. Thus, an additional force may be defined as:

$$Fx_i = RPM(x_s - x_i) \qquad (5.49)$$

where RPM is based on the appropriate cooling technology and x_s is the location next to the fluid inlet or the heat sink. The value of RPM is dependent on the cooling technology, the average of non-zero entries in the connectivity matrix, ξ, and the maximum value of the placement metric. For the convection case, RMP_i

is defined as:

$$RMP_i = \frac{CVP3_i\xi}{MAX_{j\in N}(CVP3_j)} \qquad (5.50)$$

For two-edge heat sink conduction, x_s is dependent on the module location. Modules to the right of the center line of the substrate are attracted to the right heat sink, and modules to the left are attracted to the left heat sink. The force-directed approach offers greater processing speed, and more control over the combined effects for placement, especially with the case of reliability and routability. However, a certain amount of physical reality is sacrificed in the employment of a fixed placement configuration.

5.6 References

[ALT82]   Altoz, F., et al., " A Method for Reliability-Thermal Optimization," *Proc. Ann. Reliability and Maintainability Symp.*, 1982, pp. 303-308.

[BEN81]   Bendsøe, M., " Some Smear-out Models for Stiffened Plates with Applications to Optimal Design," *Int. Symposium on Optimum Structural Design*, Univ. of Arizona, Tucson, Oct. 1981.

[BUR78]   Burroughs, J. D., et al., " Avionics/ECS CAS Study," *AFFDL-TR-78-184*, two volumes, Dec. 1978.

[CHA91]   Chang, T. S., and Magrab, E. B., "An Improved Procedure for the Determination of the Elastic Constants of Component-Lead-Board Assemblies," *ASME J. of Electronic Packaging*, vol. 113, 1991, pp. 427-430.

[CHA92]   Chang, T. S., " A Method to Increase the Fundamental Natural Frequency of Printed Wiring Board Assemblies," Ph.D. Dissertation, Univ. of Maryland, College Park, 1992.

[CHE82]   Cheng, K-T., and Olhoff, N., "Regularized Formulation for Optimal Design of Axisymmetric Plates," *Int. J. Solid Structures*, vol. 18-2, 1982, pp. 153-169.

[ENG88]   Engel, P. A., and Lim, C. K., "Stress Analysis in Electronic Packaging," *Finite Elements in Analysis and Design*, vol. 4, 1988, pp. 9-18.

[ENG90]   Engel, P. A., "Structural Analysis for Circuit Card Systems Subjected to Bending," *ASME J. of Electronic Packaging*, vol. 112, 1990, pp. 2-10.

[ENG91]   Engel, P. A., Caletka, A. V., and Palmer, R., " Stiffness and Fatigue Study for Surface Mounted Module/Lead/Card Systems," *ASME J. of Electronic Packaging*, vol. 113, 1991, pp. 129-131.

[GUY65]   Guyan, R. J., "Reduction of Stiffness and Mass Matrices," *AIAA*, vol. 3, 1965, pp. 380.

[HAK89]   Hakim, E. B., *Microelectronic Reliability, Reliability Tests and Diagnostics,* vol. I, Artech House, 1989.

[IRO65]   Irons, B. M., "Structural Eigenvalue Problem: Elimination of Unwanted Variables," *AIAA*, vol. 3, 1965, pp. 961-962.

[KID73]   Kidder, R. L., "Reduction of Structural Frequency Equations," *AIAA*, vol. 11, 1973, pp. 982.

[KID75]   Kidder, R. L., "Reply by Author to A. H. Flax, " *AIAA*, vol. 13, 1975, pp. 702-703.

[MAY78]   Mayer, A. H., " Computer Aided Thermal Design of Avionics for Optimum Reliability and Life Cycle Cost," *Technical Report AFFDL-TR-78-48,* Air Force Flight Dynamics Laboratory, 1978.

[MAY81]   Mayer, A. H., " Opportunities for Thermal Optimization in Electronics Packaging," *International Electronics Packaging Society Conference,* Nov. 1981.

[OLH81]   Olhoff, N., Lurie, K. A., Cherkaev, A. V., and Fedorov, A. V., "Sliding Regimes and Anisotropy in Optimal Design of Vibration Axisymmetric Plates," *Int. J. Solids Structures*, vol. 17-10, 1981, pp. 931-948.

[OST89]   Osterman, M., and Pecht, M., "Component Placement for Reliability on Conductivity Cooled Printed Wiring Boards," *ASME J. of Electronic Packaging*, vol. 111, June 1989, pp. 149-156.

[OST90]   Osterman, M., and Pecht, M., "Component Placement for Reliability and Routability of Convectivity Cooled PWBs," *IEEE Trans. on Computer Aided Design of Integrated Circuits and Systems*, vol. 9, July 1990, pp. 734-744.

[PAZ83]   Paz, M., "Practical Reduction of Structural Eigenproblems," *J. of Structural Division of ASCE*, vol. 109, 1983, pp. 2591-2599.

[PAZ84]   Paz, M., "Dynamic Condensation," *AIAA*, vol. 22, 1984, pp. 724-727.

[PEC87]   Pecht, M., Palmer, M., Schenke, W., and Porter, R., "An Investigation into PWB Component Placement Tradeoffs," *IEEE Trans. on Reliability*, vol. R-36, 1987, pp. 524-527.

[PEC91]   Pecht, M., *Handbook of Electronic Package Design*, Marcel Dekker, New York, 1991.

[POR78]   Porter, R. F., et al., " Integrated Thermal Avionics Design (ITAD)," *AFFDL-TR-78-76*, June 1978.

[QUI79]   Quinn, N., and Breuer, M., "A Force Directed Component Placement Procedure for Printed Circuit Boards," *IEEE Trans. CAS*, vol. CAS-26, no. 6, 1979, pp. 377-388.

[STE88]   Steinberg, D. S., *Vibration Analysis for Electronic Equipment*, John Wiley & Sons, New York, 1988.

[WOL81]   Wold, W. et al., " Integrated Thermal Avionics Design Functional Description," *AFWAL-TR-80-3148*, May 1981.

[WON90]   Wong, T. L., Stevens, K., Wang, J., and Chen, W., " Strength Analysis of Surface Mounted Assemblies under Bending and Twisting Loads," *ASME J. of Electronic Packaging*, vol. 112, 1990, pp. 168-174.

Chapter 6

Detailed Routing

Yeun Tsun Wong, Michael Pecht and Guoqing Li

A planar routing workspace, whether a hybrid or multichip module substrate or a circuit card, consists of stacks of conduction sheets sandwiched between insulation sheets (see Figure 6.1). The conduction sheet is either a ground plane, a power plane, or a signal layer. The etched space between two parallel neighboring signal paths in a signal layer is called the clearance. An electrical connection between two or more layers is called a via, and that extends through all layers is called a plated through hole.

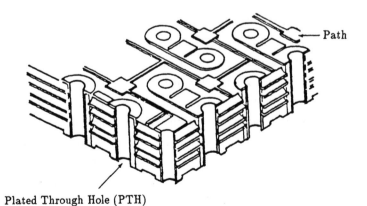

Plated Through Hole (PTH)

Figure 6.1. Cross-section of a typical substrate

The input needed to start routing includes:

- net list, that indicates all signal sets;
- workspace dimensions, number of available signal layers;
- ground and power planes;
- fan-in and fan-out of devices;
- connectors and test structures;
- dimensions on vias, routing tracks and clearances;
- requirements for special circuits.

The output from routing includes the locations of the routing paths, vias and pads on the prescribed layers.

The problem of finding or searching for a path that connects a terminal pair on a gridded workspace with obstacles is similar to the maze searching problem explored by C. Shanno in the 1940's, where a mechanical mouse was required to find its way through a checkerboard maze. In 1959, E. Moore [MOO59] presented an algorithm for maze searching, and in 1961, C. Lee [LEE61] applied Moore's algorithm to routing. In maze searching, grids are expanded to their adjacent grids, and the expanded space associated with the searching information are stored. Because the number of grids needed to search for a connection of a terminal pair was extremely large using Lee's router, especially when there were few obstacles in a high-density workspace, K. Mikami and K. Tabuchi [MIK68], D. W. Hightower [HIG69, HIG74] and F. O. Hadlock [HAD77] independently proposed routers with improved searching strategies. The searching strategy of Hadlock's router is still a maze searching technique, while Mikami-Tabuchi's and Hightower's routers are line searching techniques, which involve applying and storing straight-line segments, extended from imaginary grids occupied by other straight-line segments, to determine a path connecting a terminal pair.

The workspace for both maze searching and line searching is gridded to aid in searching and in accommodating the paths and clearances between neighboring parallel paths. Another searching technique, called gridless searching, was developed from the line searching technique, in which line segments are extended from polygon boundaries of obstacles, rather than from imaginary grids [LAU80, OHT84, ASA85, FIN85, OHT85, LUN88, SCH90, SET90].

Any area occupied by a path associated with its neighboring clearance for one net is an obstacle for another net, because two nets cannot intersect. Obstacles are expressed differently in gridded and gridless workspace. In a gridded workspace, a path and its neighboring clearance are expressed as a sequence of adjacent grids, and a grid occupied by the path and its neighboring clearance is an obstacle. In a gridless workspace, a polygon area enclosing paths and their neighboring clearance is an obstacle.

This chapter introduces various search techniques using maze searching, line searching, and gridless searching. A technique combining the advantages of three techniques, and techniques for high-speed circuitry routing are then proposed.

6.1 Maze Searching

Maze searching involves gridding the workspace to measure the length of a path by the number of grids transversed. The grid size is chosen to accommodate the path width, the clearance between parallel paths, and the vias. For a given source grid and a given target grid in a gridded workspace with some grids as obstacles, the goal of maze searching is to find a path that transverses the minimum number of adjacent available grids from the source grid to the target grid [RUB74, HOE76, SOU78, CLO84, HU85, OHT86, IOS86]. Because the search progresses from one grid to adjacent grids, maze searching is also often referred to as grid expansion.

Two search strategies are presented: Lee's router, which minimizes the number of adjacent available grids from the source grid to the target grid, and Hadlock's router, which minimizes the detour length (the number of additional grids needed to turn around obstacles). In addition, modified Lee's routers and the minimum detour-length theory are also presented.

Example: Figure 6.2a depicts a simplified gridded workspace without explicitly showing tracks and clearances, where trees t_1, t_2 and t_3 connect signal sets $V_{n_1} = \{p_1, p_2, p_3\}$, $V_{n_2} = \{p_4, p_5, p_6\}$ and $V_{n_3} = \{p_7, p_8\}$, respectively. Figures 6.2b and 6.2c depict workspaces with clearances and tracks for accommodating line segments and show a possible two-layer routing with horizontal and vertical line segments. To reduce vias, t_3 is routed in one layer (Figure 6.2c). To connect paths which belong to t_2 and are routed on different layers, via v (Steiner point s) is employed (Figures 6.2a and 6.2c). When a maze router is implemented, clearance is considered to be constant, and the simplified workspace in Figure 6.2a is used.

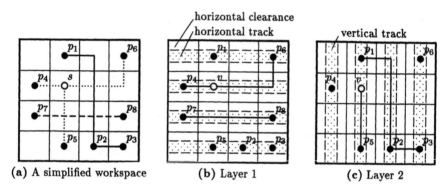

(a) A simplified workspace   (b) Layer 1   (c) Layer 2

Figure 6.2. A simplified workspace with two-layer routing

◇

6.1.1   Lee's router

The original Lee's router was employed to find a shortest path connecting a terminal pair in a single layer. Lee's router is referred to as a wave-propagation method, because the method simulates a propagating wave from a radar station.

Let the source grid, s, be the "radar station", the target grid, τ, be the "target", and the darkened grids be obstacles (Figure 6.3). Considering the slant line segments in Figure 6.3 as the wave front, the wave propagates from s to its adjacent available grids, and these grids are labeled to demonstrate that they are simultaneously scanned by the wave. The scanned grids are also considered obstacles preventing the wave from propagating back to the radar station. Next, all available grids adjacent to the grids are labeled as the wave propagates. Labeling the target grid as 0, every available grid adjacent to every grid labeled n is labeled with $n+1$ to note that the wave has propagated in time $n+1$. If the grid labeled τ is scanned in time $n+1$, then the target is reached, and scanning is terminated. If no available grid can be found, then no path exists from s to τ. This process for wave propagating from the grid s to the grid τ is called the forward search in Lee's router (Figure 6.3a).

After the target is found, the wave is reflected from the target back to the radar station along a shortest path. If the grid labeled τ was scanned in time $n+1$, then the shortest path progresses through a sequence of grids labeled n, $n-1$, ..., 2, 1 and finally, s (Figure 6.3b). Finding this path is called the backward search in Lee's router. Usually, more than one path can be found in the backward search.

Because vertical and horizontal line segments are often placed on different layers, the path with the minimum number of bends (right angles) is often selected. A path can be expressed as a sequence of adjacent points (grids) or a sequence of bend vertices.

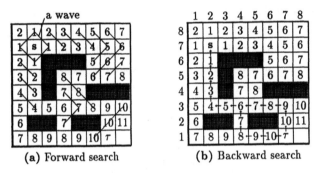

(a) Forward search   (b) Backward search

Figure 6.3. An example for Lee's router

Example: Two different paths can be found in the backward search shown in Figure 6.3b. They can be expressed as two sequences of grids: $e_{s\tau} = \{(7,1), (6,1), (5,1), (4,1), (4,2), (4,3), (3,3), (2,3), (2,4), (2,5), (2,6), (2,7)\}^*$, and $e'_{s\tau} = \{(7,1), (7,2), (7,3), (6,3), (5,3), (4,3), (3,3), (2,3), (2,4), (2,5), (2,6), (2,7)\}^*$. If a path is expressed as a sequence of bend vertices, then $e_{s\tau} = \{(7,1), (4,1), (4,3), (2,3), (2,7)\}^*$, and $e'_{s\tau} = \{(7,1), (7,3), (2,3), (2,7)\}^*$, where s=(7,1), τ=(2,7) and other points in $e_{s\tau}$ and $e'_{s\tau}$ are bend vertices.

◇

6.1.2 Modified Lee's routers

Lee's router can be modified to further reduce path length, speed up the searching process, and connect a net with more than two terminals on a two-layer workspace. Various modifications are overviewed below.

6.1.2.1 The octilinear technique

A technique modifying a rectilinear path, also called a 45° layout, to an octilinear path is often employed to reduce the path length. With this technique, a bend $\{(x, y-1), (x,y), (x+1,y)\}$, routed on the same layer, is replaced by a slant, or octilinear, line $\{(x, y-1), (x+1,y)\}$, when this octilinear line does not cross other nets. The path length is thus reduced $2 - \sqrt{2} = 0.586$. Bends that are nested one by another are simultaneously octilinearized.

Example: The routing paths shown in Figure 6.3b are modified to octilinear paths. As shown in Figure 6.4, the path connecting **s** and τ is either $e_{s\tau} = \{(7,1),(5,1),(4,2),(3,3),(2,4),(2,7)\}^*$, or $e'_{s\tau} = \{(7,1),(7,2),(6,3),(3,3),(2,4),(2,7)\}^*$. In $e_{s\tau}$, three bends are octilinearized, and the length of this octilinear path is reduced by $3 \times 0.586 = 1.758$. In $e'_{s\tau}$, two bends are octilinearized, and the length of this octilinear path is reduced by $2 \times 0.586 = 1.172$. Path $e_{s\tau}$ has the shorter length after modification.

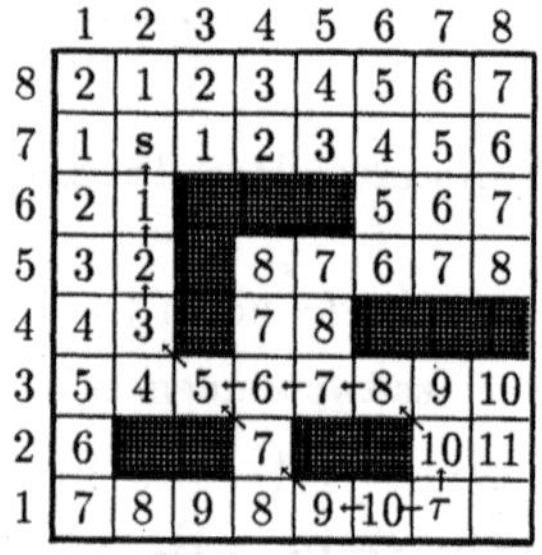

	1	2	3	4	5	6	7	8
8	2	1	2	3	4	5	6	7
7	1	s	1	2	3	4	5	6
6	2	1	▓	▓	▓	5	6	7
5	3	2	▓	8	7	6	7	8
4	4	3	▓	7	8	▓	▓	▓
3	5	4	5	6	7	8	9	10
2	6	▓	▓	7	▓	▓	10	11
1	7	8	9	8	9	10	τ	

Figure 6.4. The octilinear technique

◇

6.1.2.2 Reducing the number of scanning grids

When only a few obstacles exist in a workspace (Figure 6.5a), the number of grids that require scanning in the forward search is large. In this case, the maximum number of scanned grids, N_{maxsc1}, can be approximated by:

$$N_{maxsc1} = 1 + \sum_{i=1}^{l_{e_{s\tau}}} 4i = 1 + 2(1 + l_{e_{s\tau}})l_{e_{s\tau}} \tag{6.1}$$

where $l_{e_{s\tau}}$ is the length of path $e_{s\tau}$. Therefore, the time complexity of finding a shortest path for an edge connecting s and τ is $O(l_{e_{s\tau}}^2)$. To speed up Lee's router, the number of scanned grids can be reduced by dynamically setting a new boundary near to the source and the target grids, so that the grids outside the new boundary are not scanned (Figure 6.5b).

(a) s in the center

			3			
		3	2	3		
	3	2	1	2	3	
3	2	1	s	2	3	3
	3	2	1	2	3	
		3	2	3		
			3			

(b) s next to a boundary

3							
2	3						
1	2	3					
s	1	2	3				
1	2	3					
2	3						
3							

(c) Two waves

		3	2	3			
		3	2	1	2	3	
		2	1	s	1	2	3
		3	2	1	2	3	2	3
			3	2	3	2	1	2	3
				3	2	1	τ	1	2
					3	2	1	2	3
						3	2	3

Figure 6.5. Scanned grids in the forward search

If both the source and the target are selected as radar stations and the forward search is terminated as the two waves encounter each other (Figure 6.5c), the scanned grids can be reduced, because the number of scanned grids by one wave is proportional to the square of the path length (Equation 6.1). By using two waves, each wave scans only half of the path. In Equation 6.1, the maximum number of scanned grids, N_{maxsc2}, is

$$N_{maxsc2} = 2\left(1 + \sum_{i=1}^{l_{esr}/2} 4i\right) = 2 + (2 + l_{esr})l_{esr} \tag{6.2}$$

If the path length $l_{esr} \gg 1$, then $N_{maxsc1}/N_{maxsc2} \approx 2$. Thus, the speed for finding a path is raised almost 100%, when two waves are propagated in a workspace with few obstacles.

Example: In Figure 6.5c, waves are generated for both the source, s, and the target, τ. A path connecting s and τ consists of two sub-paths: one connects a grid labeled **3** and the grid labeled s, and another connects a grid labeled **3** and the grid labeled τ. A total of $C(6,3) = 20$ paths connecting s and τ can be selected from the rectangle, with s and τ as vertices.

◇

If grids are expanded on the perimeter of the rectangle with the source and the target as vertices to their vicinity grids, then two L-shaped spaces result, with the source and the target as end-vertices. To restrict the number of the scanned grids, Tada *et al.* [TAD80] employed a method in which each L-shaped space is iteratively expanded. If a path cannot be found in one L-shaped area, then the search starts in another L-shaped space. If a path cannot be found in both the

L-shaped spaces, then the L-shaped spaces are expanded until either a path is found or the rectangle is filled up. This modification is available only in the early routing stage.

6.1.2.3 Boolean labeling technique

If a boolean notation is used, instead of an integer notation, to indicate wave propagation, the memory required for the forward search in Lee's algorithm can be reduced. Furthermore, the additions and subtractions for indexing in both the forward and backward searches can be eliminated, so that the searching time is also reduced. S. B. Akers [AKE67] actually carried this approach one step further, proposing a labeling scheme in which instead of the series s, 1, 2, 3, ..., τ, the series s, 1, 1, 0, 0, 1, 1, ..., τ is used to label the wave propagation in the forward search. If the target grid is labeled with the first 1, then the first grid to be found in the backward search is labeled 0, and the numbers in the grids of a path are labeled τ, 0, 0, 1, 1, ..., 0, 0, 1, 1, s. However, if the target grid is labeled with the second 1, then the numbers in the grids of a path are labeled τ, 1, 0, 0, 1, 1, ..., 0, 0, 1, 1, s.

Example: Figure 6.6 shows an example of Akers' labeling modification for Figure 6.3. Note that the waves that propagate in both Lee's and Akers' approach are the same. When Akers' labeling is employed, τ is labeled with the first 0 (Figure 6.6a). Therefore, the path found in the backward search is τ, 1, 1, 0, 0, ..., 0, 0, 1, 1, s (Figure 6.6b).

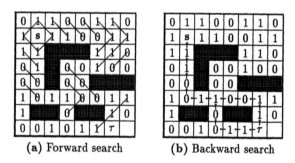

(a) Forward search   (b) Backward search

Figure 6.6. Akers' labeling modification for Lee's router

6.1.2.4 Techniques for routing a net

An approximate method for finding a tree connecting a set of terminals on a workspace involves connecting terminals one by one. If maze searching is employed, the target grid is extended to a set of grids occupied by paths for the same net. Then a new path is searched to connect a new terminal to the path.

Example: Figure 6.7 shows an example of routing a net with more than two terminals. In Figure 6.7a, the path connecting p_1, p_2 and p_4 and all grids associated with this path are considered target grids, and are thus labeled τ. In the search for a path to p_3, p_3 is considered the source grid and Lee's router can be used (Figure 6.7b). This method provides a Steiner tree arrangement.

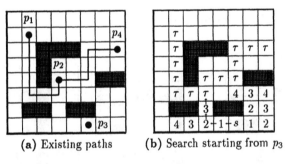

(a) Existing paths   (b) Search starting from p_3

Figure 6.7. Connecting a terminal to a sub-net

◇

Because terminals are connected one by one, a tree connected by paths found by maze searching depends on the connection order of the terminals. However, the optimal connection order pertaining to the minimum tree length is unknown. Usually, the connection order is based on the rectilinear (Manhattan) distance. For a set of terminals $V_n = \{p_1, p_2, ..., p_n\}$, if the distance between p_i and p_j $(1 \leq i \neq j \leq n)$ is the minimum (or maximum), then p_i and p_j are selected as the first and the second terminals in the connection order. Then, a terminal $p_k \in (V_n - p_i - p_j)$ that is nearest (or farest) to p_i and p_j is selected as the third terminal, and so on, until the order for every terminal in V_n is determined. In connecting a tree, the first path, $e_{p_i p_j}$, for p_i and p_j is searched, then the path for connecting p_k to path $e_{p_i p_j}$, the path for connecting the fourth terminal to the sub-tree connecting p_i, p_j and p_k, ..., are found in turn.

Example: Figure 6.8 shows the effect of the connection order on the tree. Based on the minimum distance between two terminals, the connection order is arranged as p_1, p_2, p_3 and p_4, because the distance between p_1 and p_2 is 5, the distance between p_2 and p_3 is 5, and the distance between p_4 and p_2 is 6. From the connection order, the path for connecting p_1 and p_2 is searched, then the path for connecting p_3 to the existing path is found. In this case, a Steiner point is generated. Finally, a path for connecting p_4 to the existing sub-tree is found (see Figure 6.8a). The connection order, which is arranged as p_1, p_3, p_4 and p_2, is based on the maximum distance between two terminals (see Figure 6.8b). A comparison of the trees in Figures 6.8a and 6.8b demonstrates that the tree corresponding to the connection order based on the minimum distance has a shorter tree length. However, neither the tree in Figures 8.6a, nor the tree in Figure 8.6b has the minimum tree length. The tree that dose not correspond to the previous connection orders has the minimum length (see Figure 8.6c).

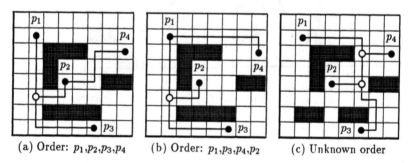

(a) Order: p_1,p_2,p_3,p_4 (b) Order: p_1,p_3,p_4,p_2 (c) Unknown order

Figure 6.8. The connection order and the resulting routing

$\diamond$

6.1.2.5 Techniques for two-layer routing

By extending a two-dimensional workspace into a three-dimensional workspace and by simultaneously expanding grids with the same horizontal and vertical coordinates onto adjacent layers, Lee's router can be modified to a multilayer router. If one of the two grids with the same horizontal and vertical coordinates in different layers is not obstructed, then the expansion can be implemented on these two grids, even though the other grid is an obstacle. However, obstacles labeled in the forward search cannot be collected to a path in the backward search.

Example: Figure 6.9 shows an example of a search on two adjacent layers. Figure 6.9a shows obstacles and paths for an existing net, in which the solid line and the dotted line represent paths on layer 1 and layer 2, respectively. The obstacles formed by the path shown in Figure 6.9a are included in Figures 6.9b and 6.9c. The terminals and the via are obstacles on both the layers. The wave is simultaneously propagated on both the layers from the source, s, to the target, τ, through each pair of grids with the same coordinate and with at least one available. The grids in Figure 6.9b with the coordinates as those of grids labeled 4, 5, 6 and 7 in row 3 in Figure 6.9c are obstacles, however, they are labeled as if they are available. In Figures 6.9b and 6.9c, three paths can be found on the same layer: $\{(7,1,1), (7,3,1), (6,3,1), (6,7,1), (2,7,1)\}^*$ in layer 1, and $\{(7,1,2), (4,1,2), (4,3,2), (2,3,2), (2,7,2)\}^*$ and $\{(7,1,2), (7,3,2), (2,3,2), (2,7,2)\}^*$ in layer 2. Paths occupying grids on two layers are $\{(7,1,1), (7,3,1), (7,3,2), (2,3,2), (2,7,2)\}^*$, $\{(7,1,1), (7,3,1), (7,3,2), (2,3,2), (2,4,2), (2,4,1), (2,7,1)\}^*$, and so forth.

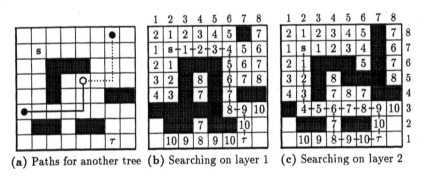

(a) Paths for another tree   (b) Searching on layer 1   (c) Searching on layer 2

Figure 6.9. Searching on two adjacent layers

Only considering those obstacles obstructing grids with the same coordinates in both adjacent layers in the forward search, the two-layer workspace can be conducted to a single-layer workspace, so that implementing Lee's router separately in two layers can be avoided. If the path is routed through a grid occupied by an obstacle on one layer, then this path must be assigned to the grid in the other layer in the backward search.

Example: Because only those obstructed grids with the same coordinates in both layers are considered as obstacles, the grids occupied by the path shown in Figure 6.9a are not considered as obstacles, except for two terminals and the via.

The labeling process of the forward search in the single workspace is the same as that shown in Figure 6.9b or 6.9c. Therefore, the paths found by the two-layer workspace can be searched in the single-layer workspace.

◇

6.1.3 Minimum detour-length searching

Let the length of a path $e_{s\tau}$ connecting the source **s** and the target τ be $l_{e_{s\tau}}$. Considering obstacles, $l_{e_{s\tau}}$ can be expressed as

$$l_{e_{s\tau}} = |x_s - x_\tau| + |y_s - y_\tau| + L_{total\_detour} \qquad (6.3)$$

where L_{total_detour} is the total detour length, which is required for routing around obstacles. Because the rectilinear distance $|x_s - x_\tau| + |y_s - y_\tau|$ is a constant, minimizing L_{total_detour} is equivalent to minimizing $l_{e_{s\tau}}$. In order to develop minimum detour-length techniques, the minimum detour length theory is presented. Then, Hadlock's router based on this theory is overviewed.

6.1.3.1 Minimum detour-length theory

Let path $e_{s\tau} = \{s, p_1, p_2, ..., p_n, \tau\}^*$ be composed of the straight line segments e_{sp_1}, $e_{p_i p_{i+1}}$ $(i = 1, ..., n-1)$ and $e_{p_n\tau}$. When the line segment, $e_{p_i p_{i+1}}$, from p_i to p_{i+1}, cannot be enclosed completely by the rectangle with p_i and τ as vertices, the grids outside the rectangle form a detour from p_i to p_{i+1} (or a detour in $e_{p_i p_{i+1}}$). The number of grids in the detour from p_i to p_{i+1} is called the detour length of this detour, $l_{detour}(p_i, p_{i+1})$, that can be computed by:

FUNCTION $l_{detour}(p_i, p_{i+1})$;
BEGIN
    IF $x_{p_i} = x_{p_{i+1}}$ THEN BEGIN
        IF $y_{p_i} \leq y_\tau \leq y_{p_{i+1}}$ OR $y_{p_{i+1}} \leq y_\tau \leq y_{p_i}$ THEN
            $l_{detour} = |y_{p_{i+1}} - y_\tau|$
        ELSE IF $y_{p_{i+1}} \leq y_{p_i} \leq y_\tau$ OR $y_\tau \leq y_{p_i} \leq y_{i+1}$ THEN
            $l_{detour} = |y_{p_i} - y_{p_{i+1}}|$
    END ELSE BEGIN
        IF $x_{p_i} \leq x_\tau \leq x_{p_{i+1}}$ OR $x_{p_{i+1}} \leq x_\tau \leq x_{p_i}$ THEN
            $l_{detour} = |x_{p_{i+1}} - x_\tau|$

$$ELSE \; IF \; x_{p_{i+1}} \le x_{p_i} \le x_\tau \; OR \; x_\tau \le x_{p_i} \le x_{i+1} \; THEN$$
$$l_{detour} = |x_{p_i} - x_{p_{i+1}}|$$
$$END;$$
$$END.$$

If detour e_1 starts from a grid in the perimeter of the rectangle with s and τ as vertices, either a parallel detour with the length of e_1 exists, that terminates at another grid in the perimeter, or the path from s to τ, that includes detour e_1, is broken. Corresponding to every detour length computed by *FUNCTION* l_{detour}, there is a parallel detour with the same length. However, this length is zero if it is computed by the function. Considering the detour length corresponding to the detour in $e_{p_i p_{i+1}}$ as well as $l_{detour}(p_{n-1}, p_n) = 0$ for $p_n = \tau$, the detour length from s $= p_0$ to τ, $L_{total\_detour} = 2L_{detour}$, where:

$$L_{detour} = \sum_{i=0}^{n-2} l_{detour}(p_i, p_{i+1}) \tag{6.4}$$

and the length of path $e_{s\tau}$ is computed by:

$$l_{e_{s\tau}} = |x_s - x_\tau| + |y_s - y_\tau| + 2L_{detour}. \tag{6.5}$$

Equations 6.4 and 6.5 show that minimizing $l_{detour}(p_i, p_{i+1})$ ($i = 1, ..., n - 2$) is equivalent to minimizing $l_{e_{s\tau}}$.

Example: Figure 6.10 shows the determination of detours and their lengths in path $e_{s\tau} = \{s, p_1, ..., p_5, \tau\} = \{(3,3), (5,3), (5,9), (7,9), (7,4), (11,4), (11,6)\}^*$. Because p_1 is enclosed by the rectangle with s and τ as vertices (see the dash line in Figure 6.10), using e_{sp_1} as a section of path, $e_{s\tau}$, will not increase the path length. The detour length from s to p_1, that can be computed by *FUNCTION* l_{detour}, is $l_{detour}(s, p_1) = 0$. Considering p_1 as a new source, p_2 is not enclosed by the rectangle with p_1 and τ as vertices. Using $e_{p_1 p_2}$ as a section of path, $e_{p_1 \tau}$, the path length will increase from $l_{e_{p_1 \tau}} = |x_{p_1} - x_\tau| + |y_{p_1} - y_\tau|$ to $l_{e_{p_1 \tau}} + 2 \times l_{p'_1 p_2}$, because detours $e_{p'_1 p_2}$ and $e_{p_3 p'_3}$ are needed for turning around the obstacles. Calling Function l_{detour}, $l_{detour}(p_1, p_2) = l_{e_{p'_1 p_1}} = 3$, and $l_{detour}(p_3, p_4) = l_{e_{p'_3 p_4}} = 2$. Note that detour length $l_{detour}(p_1, p_2) + l_{detour}(p_3, p_4) = l_{detour}(p'_1, p_2) + l_{detour}(p'_3, p_4) = 5$ is the half-length of the detours needed for turning around the obstacles, because $e_{p_3 p'_3}$ and $e_{p_5 \tau}$ are associated with detours $e_{p'_1 p_2}$ and $e_{p'_3 p_4}$, respectively, and $l_{detour}(p_2, p_3) = l_{detour}(p_3, p'_3) = l_{detour}(p_4, p_5) = l_{detour}(p_5, \tau) = 0$.

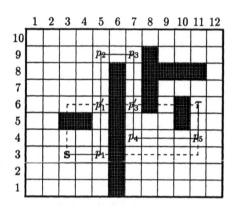

Figure 6.10. Detours and detour length

◇

An iso-detour octagon is composed of grids, whose distances to grids in the perimeter of the rectangle, with two terminal grids as vertices, equal a specific detour length. When a path is required to be longer than the rectilinear distance between two terminals by a specific length, the path can be found by connecting the terminal grids either to two grids on the same horizontal or vertical line, or to a grid on a octilinear (slant) line, with the rectilinear distance, in the iso-distance octagon [TAD86, HAN90].

Example: Figure 6.11 shows an iso-detour octagon with a dot line for detour length equal to 2. Connecting s, p_1, p_1' and τ or connecting s, p_2 and τ, a path with a total detour length equal to 4 is obtained.

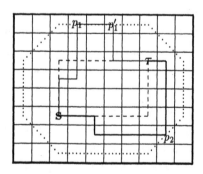

Figure 6.11. An iso-detour octagon for the detour length equal to 4

◇

6.1.3.2 Hadlock's router

In order to minimize the detour length from one grid to its adjacent grid, F. O. Hadlock [HAD77] developed a minimum detour-length algorithm to search for the shortest paths between two grids. Hadlock's router is a special case of the minimum detour-length technique. In Hadlock's router, as with Lee's router, there is both a forward search and a backward search. In the forward search, if available grids p_{i+1}, p'_{i+1}, ..., and $p_{i+1}^{(n)}$ ($n \leq 3$) are adjacent to p_i labeled n_{p_i}, and $l_{detour}(p_i, p_{i+1})$ is equal to or less than $l_{detour}(p_i, p_{i+1}^{(i)})$ ($i = 1, ..., n$), then grid p_1 is expanded to p_{i+1}, and p_{i+1} is labeled $n_{p_i} + l_{detour}(p_i, p_{i+1})$, where $l_{detour}(p_i, p_{i+1}) = 0$ or 1.

Initially, the source grid, s, is labeled 0. Grid p_1 adjacent to s is labeled with 0 if $l_{detour}(s, p_1) = 0$. After labeling all grids with detour-length equal to 0, grids with a detour-length equal to 1 are labeled, and the process continues until τ is labeled.

Hadlock's router can find all shortest paths found by Lee's router. The time complexity for an $N \times N$ workspace ranges from $O(N)$ to $O(N^2)$, depending on the locations of the source grid, the target grid and obstacles. If there are few bends in the path, then Hadlock's router runs faster than Lee's router in finding the shortest paths.

Example: Figure 6.12 shows the forward search in Hadlock's router. In this Figure, s=(3,3) and $\tau = (11,6)$. Grids (3,4) and (4,3) are labeled 0, because they are enclosed by the rectangle with s and τ as vertices. Grids (4,4) and (5,3) are labeled 0, because (4,4) is enclosed by the rectangle with (3,4) and τ as vertices, and (5,3) is enclosed by the rectangle with (4,3) and τ as vertices. Similarly, (5,4), (5,5) and (5,6) are labeled 0. Because (4,6) and (5,7) are adjacent to (5,6) labeled 0 and they are not enclosed by the rectangle with (5,6) and τ as vertices, (4,6) and (5,7) are labeled 1. After all grids with detour length equal to 1 are labeled, all grids adjacent to grids labeled 1 that have detour length equal to 1 are labeled 2. Note that grid (5,9) is not enclosed by the rectangle with (5,8) and τ as vertices, and is labeled 3. Because grid (6,9) is enclosed by the rectangle with (5,9) and τ as vertices, (7,9) is enclosed by the rectangle with (6,9) and τ as vertices, and (7,8) is enclosed by the rectangle with (7,9) and τ as vertices, etc., grids (6,9), (7,9), (7,8), (7,7) and (7,6) are labeled 3. Finally, τ is labeled 5.

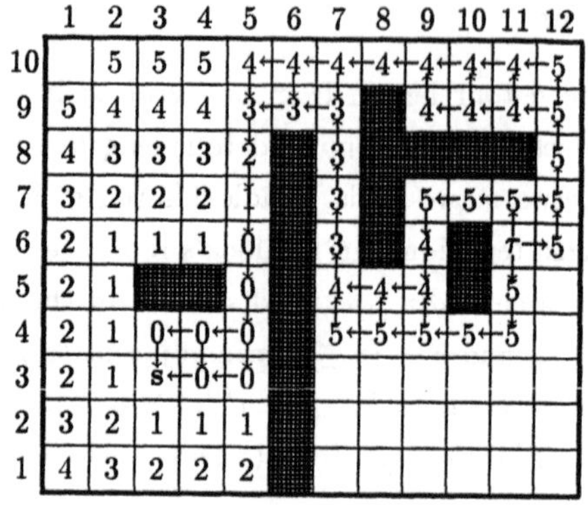

	1	2	3	4	5	6	7	8	9	10	11	12
10		5	5	5	4←	4←	4←	4←	4←	4←	4←	5
9	5	4	4	4	3←	3←	3	■	4←	4←	4←	5
8	4	3	3	3	2	■	3	■				5
7	3	2	2	2	1	■	3	5←	5←	5←	5	
6	2	1	1	1	0	■	3	4	■	↑→	5	
5	2	1	■		0	■	4←	4←	4		5	
4	2	1	0←	0←	0	■	5←	5←	5←	5←	5	
3	2	1	s←	0←	0	■						
2	3	2	1	1	1	■						
1	4	3	2	2	2	■						

Figure 6.12. Hadlock's router

◇

The backward search in Hadlock's router is similar to that in Lee's router. If the target grid is labeled n, the backward search examines grids labeled n to find all couples of adjacent grids that are labeled n and $n-1$. Each grid labeled $n-1$ is a child of its neighboring grid labeled n in a connection tree. Then, grids labeled $n-1$ are examined to find grids next to grids labeled $n-2$. A grid labeled $n-2$ is a child of its neighboring grid labeled $n-1$. This process continues until grids labeled 0 are found. All paths with the same length can be found from the connection tree.

Example: The connection tree generated in the backward search is shown in Figure 6.12, where τ is labeled 5 and the arrows indicate the backward search directions. Examining grids labeled 5, couples of adjacent grids labeled 5 and 4 are found to be $\{(9,7),(9,6)\}$, $\{(9,4),(9,5)\}$, $\{(8,4),(8,5)\}$, $\{(7,4),(7,5)\}$, $\{(12,9),(11,9)\}$, and $\{(12,10),(11,10)\}$. They are also stored in a connection tree as the children or grand-children of τ (Figure 6.13). Next, examining grids labeled 4 from (9,6), (9,5), (8,5) and (7,5), grid (7,6) labeled 3 is the nearest, and is stored as a child of (9,6), (9,5), (8,5) or (7,5). Note that (7,6) cannot be a child of (11,9) or (11,10), because (11,9) and (7,5) are separated by grids labeled 3 or 5. Examining grids labeled 4 from (11,9) and (11,10), grids (5,9), (6,9) and (7,9) are children of (11,9) or (11,10). Continuing the backward search, grid (5,8) is a child of (5,9), (6,9), (7,9) or (7,6), and grid (5,7) is a child of (5,8). The connection tree is shown in Figure 6.13. Ten paths can be found from this connection tree.

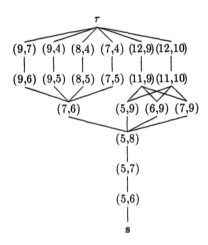

Figure 6.13. A connection tree generated in the backward search of Hadlock's router
◇

6.2 Line Searching

Instead of employing grids, the line-searching technique uses trial lines to generate
a routing path. Initially, the trial lines are extended from two terminals along
the horizontal and vertical directions and terminated by the boundary of the
workspace or obstacles. Except for those points that have generated the trial
lines, each point of a trial line can be treated as a trial point for generating a new
trial line. Because each trial point corresponds to the center of a grid, there is an
imaginary gridded workspace corresponding to the obstacles and trial lines. By
iteratively extending trial points to trial lines, and storing the trial lines onto a
search tree, an intersection can be found. The final path is found from the search
tree by searching backward from the intersection to the terminals.

6.2.1 Mikami-Tabuchi's router

In Mikami-Tabuchi's router, there exists a trial line for every trial point. Trial
lines extended from trial points of an ith-level trial line are called $(i+1)$th-level
trial lines. Initially, terminals s and τ are selected as trial points in the forward
search, and the horizontal and vertical trial lines extending from s and τ are
the 0th level trial lines. In each searching step, an ith level trial line originated

from s is stored as a parent, and every $(i+1)$th level trial line extended from every trial point in the ith level trial line is stored as children and is examined to determine whether an intersection with a trial line originating from τ exists. If an intersection is found, then a path that connects s and τ exists. Otherwise, the process is repeated.

When the intersection is found in an $(i+1)$th-level trial line, the trial points in each trial lines of level i, level $i-1$, ..., level 1, and level 0 can be found from the search tree in the backward search. In general, there are $i+j+1$ trial points (bend vertices) in the path if the intersection is generated by both an ith-level trial line originating from s and a jth-level trial line originating from τ.

Example: Figure 6.14 shows the implementation of Mikami-Tabuchi's router. Initially, four 0th-level trial lines are extended from s and τ. Because no intersection exists between the 0th-level trial lines extended from s and the 0th-level trial lines extended from τ, the 0th-level trial lines are stored as parents of the subsequent first-level trial lines in a search tree. There are ten trial points in the 0th-level trial lines extended from s, and four trial points in the 0th-level trial lines extended from τ. Because ten first-level trial lines that originate from s do not intersect with any of two 0th-level trial lines and four first-level trial lines that originate from τ, all of the first-level trial lines are stored, and the second-level trial lines are extended. Because there is an intersection between a vertical second-level trial line that originates from s and a horizontal second-level trial line that originates from τ, respectively, the forward search is completed.

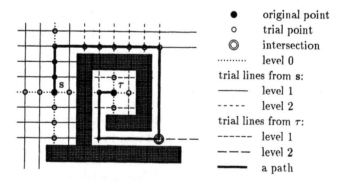

● original point
o trial point
◎ intersection
.... level 0
trial lines from s:
——— level 1
- - - - - level 2
trial lines from τ:
- - - - - - level 1
— — — level 2
——— a path

Figure 6.14. Mikami-Tabuchi's router

In the backward search, five trial points for the path connecting s and τ can be found from the search tree. Note that the second level trial line that originates from s encounters two first-level trial lines in the backward search. The trial line next to the obstacle is selected, because the trial point nearest to the intersection is in this trial line.

◇

Mikami-Tabuchi's router guarantees to find a path if it exists. If there exist more than one intersection between a trial line originating from terminal s and a trial line originating from terminal τ, not all generated paths are the shortest.

Example: Using the same obstacles as in Figure 6.3a, Mikami-Tabuchi's router generates a path crossing intersection i_1 (see Figure 6.15). However, this path is not the shortest. The shortest path can be found from other first-level trial lines.

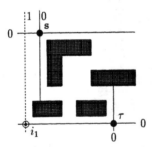

Figure 6.15. The path may not be the shortest

◇

6.2.2 Hightower's router

Instead of generating all trial lines perpendicular to an parent trial line, in Hightower's router, only a so-called escape line is considered, that is the longest over all trial lines extended from trial points in the same parent trial line, and is nearest to the trial point generating the parent trial line. Hightower's router runs faster than both Lee's and Mikami's routers, because only one escape line is extended from another escape line. Though the path found by Hightower's router has the minimum number of bends, it may not be the shortest. On average, the total path length generated by Hightower's router is less then 10% over that generated by Lee's router [HIG74]. Furthermore, in some cases with complex obstacles, an existing path may not be found by Hightower's router.

Example: Figure 6.16 shows the escape lines generated by Hightower's router for the same workspace in Figure 6.14. From the horizontal and vertical escape lines originating from terminal **s** (see Figure 6.16), a horizontal first-level escape line is generated over **s**, and a vertical first-level escape line is generated on the left of **s**. One horizontal and one vertical first-level escape lines are generated under τ and on the left of τ, respectively. However, the child trial line of the horizontal first-level escape line is the vertical first-level escape line. Therefore, the horizontal first-level escape line is not considered further. An intersection is obtained from both the vertical second-level escape line, that originates from **s**, and from the horizontal second-level escape line, that originates from τ.

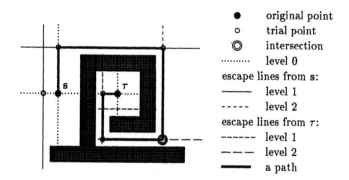

Figure 6.16. Hightower's router

◇

6.2.3   The line-expansion router

To combine the best of the maze searching and the line searching, W. Heyns *et al.* [HEY80] proposed a line-expansion router, in which a line segment called an expanded line is moved (or expanded) along its perpendicular direction. The area occupied by grids swept by an expanded line is called an expansion zone. The boundary segments of the expansion zone that are perpendicular to the expanded line are called active lines, and can be treated as expanded lines for the further expansion outside the zone. The search process in the expansion zone is similar to that in a modified Lee's router, which uses all grids on an expansion line as source grids. However, only active lines are stored for the backward search.

Example: Figure 6.17 shows an expansion zone, in which l is an expanded line, and l', l'', and l''' are active lines generated by upwardly expanding l. The labeling numbers in the figure show the waves propagating upward from source grids occupied by l. The expansion zone can also be considered as a zone swept upward by l. When l encounters the obstacle, active line l' is generated by the left end-vertex of l, and l is cut by the obstacle. Then l continues to sweep upward, and l'' and l''' are generated by the end-vertex of l. Subsequently, active lines are treated as expanded lines and are expanded along the directions of the arrow.

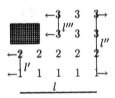

Figure 6.17. Line l and its upward expansion zone

◇

   Initially, the expansion zones are generated both by the active lines that cross the source and by the target in the forward search. The active lines of the newly generated zones are then treated as expanded lines. The process continues until the expansion zones that originated from the source and the target overlap. The overlapping zone is called a solution zone. To prevent duplication in the search process, a stop line is generated when two expanded zones originating from the same original grid, s or τ, encounter each other. Active lines generated by an expanded line in the forward search are stored in a search tree as children of the expanded line. A path is found by searching from the active lines in the solution zone back to those expanded lines extended from the source and the target.

   An existing path that connects s and τ can always be found by the line-expansion router. Because the number of grids in an expansion zone is smaller than that in the workspace for Lee's router, less memory is used. The backward search of the line-expansion router is also faster than Lee's router, because the searching transverses active lines, instead of grids.

Example: Figure 6.18 shows an example of line-expansion routing, in which the forward search in an expansion zone has been given in Figure 6.17. Figure 6.18a shows the zone-to-zone expansion from s and τ. The third expanded zone originating from s encounter with the third expanded zone originating from τ.

Note that a stop line is generated between expanded zones 2' and 3", because both the expanded zones 2' and 3" originate from τ and coincide. The path that connects s and τ is obtained from the backward search along the active lines in each expansion zone (Figure 6.18b).

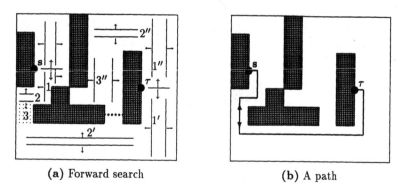

(a) Forward search   (b) A path

Figure 6.18. Line-expansion process

::::: Solution zone   ••••• Stop line

6.3 Gridless Searching

Unlike maze-searching or line-searching routers, which use a gridded workspace or an imaginary gridded workspace, a gridless router uses a workspace with available polygonal areas to accommodate new paths. Obstacles are treated as non-available polygonal areas. In gridless searching, a path consists of line segments determined by the coordinates of the boundaries of the polygonal obstacles in the gridless workspace. After a path is found, polygonal obstacles are expanded to the area occupied by the path and its associated clearance. Gridless searching techniques are especially significant for high-density routing, because the density can be ignored, and the path width can be variable. This section introduces a rectangle-expansion technique and Ohtsuki's gridless searching techniques.

6.3.1 Rectangle expansion router

If an expanded line used in the line expansion router is moved, until the boundary of the workspace or obstacles are encountered, the area swept by the expanded line is a rectangle. A path can then be found from the edges of the rectangles expanded from the source to the target. Because the rectangle expanded from an expanded line depends on both the coordinates of the boundary of the obstacles and the workspace, rather than grids, the line-expansion router is thus transformed into a gridless router. This gridless router, a rectangle expansion router, was first suggested by U. Lauther [LAU80] for developing a new data structure to improve Hightower's router.

According to the horizontal and vertical directions, line segments on the boundaries of all obstacles and the workspace can be sorted. A rectangle enclosing the source, s, with the nearest line segments can be determined by the coordinates of two nearest parallel line segments that sandwich s. Next, a new rectangle is determined by an expandable line segment, that is a section of the perimeter of the rectangle enclosing s and does not overlap with obstacles or the boundary of the workspace, and by its nearest parallel sorted line segment, that does not overlap with the previous rectangle. The process of expanding a rectangle to other rectangles from expandable line segments is continued until the target, τ, is enclosed by a rectangle. To simplify the expansions from two edges of a bend, a bend can be expanded to the same rectangle. Considering the source and the target to be two special rectangles, a rectangle is stored as the parent, and rectangles expanded from this rectangle are stored as the children in the search tree for backtracing.

In the backward search, first selects a point nearest to τ in a common line segment between the rectangle enclosing τ and its parent rectangle. Next selects a point nearest to the newly selected point in a common line segment between the rectangle enclosing the newly selected point and its parent rectangle. The similar process is continued until a point in a child rectangle of s is selected. Sequentially connecting the source, all the points selected in the backward search, and the target with the rectilinear distance, a path is obtained.

Example: An application of the rectangle expansion router is depicted in Figure 6.19. Figure 6.19a shows the obstacles in the workspace. Initially, s is expanded to rectangle R_0, in which there are three expandable line segments, $e_{p_1p_2}$, $e_{p_2p_3}$ and $e_{q_1q_2}$, which are not next to the obstacles. Here, $e_{p_1p_2}$ and $e_{p_2p_3}$ are two edges of a right-angle (bend) (Figure 6.19b). R_0 is then expanded to R_1 from $e_{p_1p_2}$ and $e_{p_2p_3}$, and to R_1' from $e_{q_1q_2}$. Because the perimeter of R_1' overlaps the boundary of the workspace and an obstacle, no rectangle can be further expanded from R_1'. In the following process, R_1 is expanded to R_2 from line segment $e_{p_4 p_5}$, R_2 is expanded to R_3 from $e_{p_6 p_7}$, and R_3 is expanded to R_4 from $e_{p_8 p_9}$. The expansion terminates in R_4, which encloses τ.

In the backward search, a point in $e_{p_8 p_9}$ and nearest to τ is selected. Denoting this point p_{89} (Figure 6.19c), p_6 is found to be a point in $e_{p_6 p_7}$ nearest to p_{89}. Similarly, a point in $e_{p_4 p_5}$ and nearest to p_6 is selected and denoted $p_{p_4 p_5}$. Because R_1 is expanded from two line segments, the point in these line segments and nearest to p_{45} should be selected. In this case, it is p_3 . Because s is enclosed by R_0, the backward search is terminated. The path is shown in Figure 6.19c.

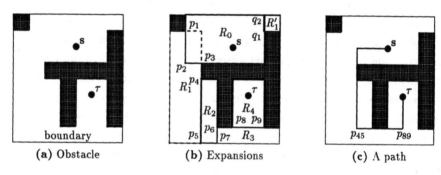

Figure 6.19. An example of the rectangle expansion router

◇

6.3.2   Ohtsuki's gridless router

In the rectangle-expansion router, an expanded rectangle consists of line segments extended from the non-concave line segments of boundaries of obstacles. The shortest path can be found from these extension line segments.

In Ohtsuki's gridless router, a set of horizontal extension lines and and a set of vertical extension lines are generated by extending each non-concave line segment until an obstacle or the boundary of the workspace is encountered. Similarly, two terminal extension lines can be generated by extending a terminal horizontally and vertically. To search for an extension line efficiently, the set of horizontal extension lines and the set of vertical extension lines are sorted in terms of their respective coordinates. Ohtsuki's gridless router is more efficient than the rectangle-expansion router, because each expandable line segment in the perimeter of an expanding rectangle must be determined for expanding one rectangle to other rectangles in the rectangle-expansion router.

Example: Figure 6.20a shows a gridless workspace, in which the dark area is an obstacle. Figure 6.20b shows the horizontal terminal extension lines and extension lines extended from all non-concave horizontal line segments of the polygonal boundary. Note that a concave line segment by τ can not be extended. Figure 6.20c shows the vertical terminal extension lines and extension lines extended from all vertical line segments of the polygonal boundary.

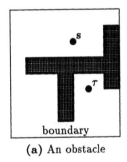
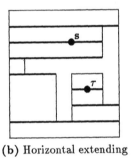
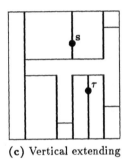

    (a) An obstacle   (b) Horizontal extending   (c) Vertical extending

Figure 6.20. Extension lines

◇

    In the forward search, the extension lines that intersect the source extension lines are called the first level extension lines, and stored in a search tree. In general, the extension lines encountered by ith-level extended lines are called the $(i+1)$th-level extension lines, and stored as the children of the ith-level extension line. If an $(i+1)$th-level extension line is a target extension line, then the forward search is completed.

    Because the length of the line segment is not considered in each step of the forward search, the path obtained in the backward search has the minimum number of bends, rather than the minimum path length. If the minimum path length is desired, all extension lines should be extended to τ, and all possible paths should be investigated in the backward search. In this way, a path with the minimum length can be obtained.

Example: Figure 6.21 shows the forward search in a gridless workspace from extension line l_{ij} to $l_{(i+1)j}$ until a target extension line is encountered, where i is a level index. Initially, source extension lines l_{01} and l_{02} are selected. As shown in Figure 6.21a, extension lines l_{11} and l_{12} are crossed by l_{01}, and l_{13} and l_{14} are crossed by l_{02}. Extension lines l_{11}, l_{12}, l_{13}, and l_{14} are stored in the search tree as the first level extension lines (see Figure 6.22). Because extension lines l_{21} and l_{22} are crossed by l_{11} (see Figure 6.21b), l_{21} and l_{22} are stored as the children of l_{11}. Because l_{13} and l_{14} are also crossed by l_{12} and the levels of l_{13} and l_{14} are not higher than l_{12}, the search stops in the branch of l_{12}. Similarly, the search stops in l_{13} and l_{14}, because l_{13} and l_{14} are also crossed by l_{11}, l_{02} and l_{12} whose levels are equal to or higher than the levels of l_{13} and l_{14}. Further searching is shown in Figure 6.21c, and the complete search tree is given in Figure 6.22. Note that l_{33} is a vertical target-extension line.

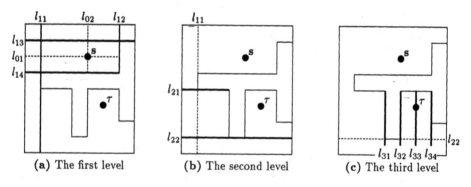

(a) The first level   (b) The second level   (c) The third level

Figure 6.21. The forward search in a gridless workspace

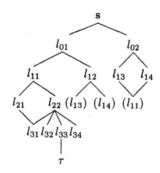

Figure 6.22. The search tree

The backward search starts from a target extension line, and traces back to its parent extension line. The parent extension line for every transversed extension line is transversed until a source extension line is encountered. An extension line transversed in the backward search must be cut both by its child extension line (or τ) and by its parent extension line (or s), so that line segments not belonging to the path are removed. If the extension lines transversed in the backward search are $l_{na}, l_{(n-1)a'}, ...,$ and $l_{0a^{(n)}}$, then the path consists of segments obtained by cutting l_{na} by τ and $l_{(n-1)a'}$, cutting $l_{(n-2)a''}$ by $l_{(n-1)a'}$ and $l_{(n-3)a'''}$, cutting $l_{(n-i)a^{(i)}}$ by $l_{(n-i+1)a^{(i-1)}}$ and $l_{(n-i-1)a^{(i+1)}}, ...,$ and cutting $l_{0a^{(n)}}$ by $l_{1a^{(n-1)}}$ and τ. If two or more parents exists for an extension line, more paths can be found.

Example: From the search tree in Figure 6.22, a sequence of extension lines, l_{33}, l_{22}, l_{11}, and l_{01}, can be found in the backward search. The path shown in Figure 6.23 consists of a section of l_{33} between τ and l_{22}, a section of l_{22} between l_{33} and l_{11}, a section of l_{11} between l_{22} and l_{01}, and a section of l_{01} between l_{11} and s.

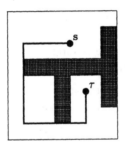

Figure 6.23. The path

◇

6.3.3   Multilayer gridless router

The one-layer gridless router can be employed for multilayer routing by placing horizontal and vertical line segments on separate layers. Before searching, all non-concave line segments of the boundary of each obstacle in a workspace for a layer accommodating horizontal (vertical) line segments are horizontally (vertically) extended. A workspace similar to that for the one-layer gridless searching can be obtained by overlaying all workspaces together.

Example: Figure 6.24 shows an example of extension lines for two-layer gridless searching with different obstacles in different layers. An obstacle, formed by horizontal routed line segments and reserved area, and horizontal extension lines are shown in Figure 6.24a. Two obstacles, formed by vertical routed line segments, and vertical extension lines are shown in Figure 6.24b. The workspace shown in Figure 6.24c is obtained by superposing extension lines and obstacles, shown in Figures 6.24a and 6.24b.

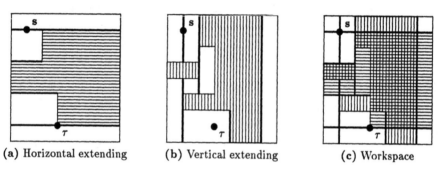

(a) Horizontal extending   (b) Vertical extending   (c) Workspace

Figure 6.24. Workspaces and extension lines in two-layer gridless searching

◇

The physical terminal structure is often restricted to a specific layer, as is the case with some surface-mount techniques, or to all layers, as is the case with plated through-holes. After the source and target extension lines are determined, the searching process is similar to that in Ohtsuki's single-layer router. However, Ohtsuki's two-layer gridless searching does not guarantee the determination of the shortest path, if the forward search stops as soon as a target extension line is encountered.

Example: Figure 6.25 shows an example of multilayer gridless searching, in which s is in common with all layers, and τ is on a layer for horizontal line segments. The forward search in the workspace starts either from the layer for horizontal line segments, or from the vertical line segments. Starting the forward search from the horizontal source extension line shown in Figure 6.25a, the resulting search tree is shown in Figure 6.25b, and the path found by the backward search is shown in Figure 6.25c. Starting the forward search from the vertical source extension line shown in Figure 6.25d, the resulting search tree is shown in Figure 6.25e, and the path found by the backward search is shown in Figure 6.25f. Path 1 in Figure

6.25c has two bends. Path 2 in Figure 6.25f has five bends and a shorter length than that of path 1. Therefore, starting from target extension lines on different layers can result in different paths. If the forward search starts from both source extension lines and stops as soon as a target extension line is encountered, then Figure 6.25c results.

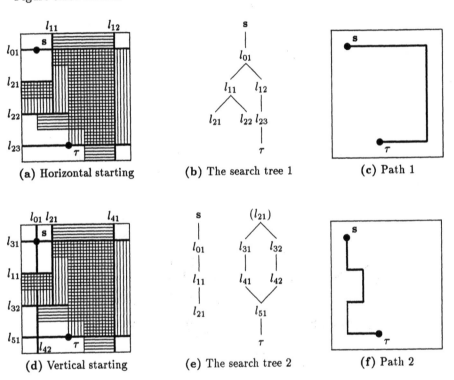

(a) Horizontal starting   (b) The search tree 1   (c) Path 1

(d) Vertical starting   (e) The search tree 2   (f) Path 2

Figure 6.25. Two-layer gridless searching

6.4   Advanced Search Techniques

Routers must be capable of addressing the effects of plated through-holes, blind vias and terminals for high-speed and multilayer circuit requirements. This section addresses multilayer searching techniques which consider vias and congestion (area costs). A gridless searching technique incorporating the minimum detour length theory, and searching techniques for special requirements are also presented.

6.4.1   Maze searching in a costing workspace

To restrict paths from passing through a "congested" area, a cost associated with
the routing density can be given to the grids in the congested area. The highest
routing density is related to the number of layers and vias and, therefore, related
to manufacturing cost. The cost associated with congestion can also be measured
by the routing length. An obstacle always has an infinitive cost because a path
cannot pass through. Thus, searching a path with the minimum cost can be
treated as searching a path with the minimum routing length.

Example: Figure 6.26 shows an example of routing on a costing workspace. In
Figure 6.26a, a grid with two numbers has double the cost of that for a regular
grid. Because of the additional grid cost in the grid with two numbers, the grid
is considered to be two grids in the forward search shown in Figure 6.26b. The
minimum cost (or routing length) of a path connecting **s** and τ in this example is
13, while the path length is 11 or 13.

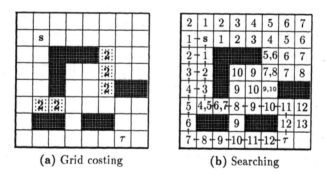

(a) Grid costing   (b) Searching

Figure 6.26. Routing on a costing workspace

◇

Often, a via is generated in the forward search, because it cannot be determined
before searching. An additional cost for a via can be assigned to a grid that is
expanded from one layer to another layer.

Example: Figure 6.27 shows grid costing for vias. In the workspace in Figure
6.27a, **s** and τ are located on layer 1. The grids darkened with horizontal line
segments are obstacles on layer 1, and the dark grids are obstacles on both layers.
The search depicted in Figure 6.27b has the additional cost of 1 for a via. When

the wave propagates from grid (2,4) to grid (2,3), it transverses from layer 1 to layer 2. If a path occupies grids (2,3) and (2,4), then there must exist a via in grid (2,3) or (2,4). In this case, grid (2,3) is scanned twice. Similarly, when the "wave" propagates from grid (4,3) to grid (4,2), it transverses from layer 2 to layer 1. A via, therefore, exists in grid (4,2) or (4,3). In this case, grid (4,2) is scanned twice. When the additional cost for vias is considered, two paths with the minimum cost results in this example: one with two vias and another without vias.

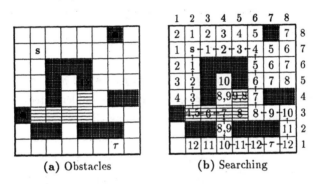

(a) Obstacles   (b) Searching

Figure 6.27. Costing a via in the forward search

Figures 6.26 and 6.27 also show that costs used in a maze searching technique are treated as integers. Because both the via costing and area costing cannot be flexibly treated as real numbers, a more complicative maze searching, that involves in re-dividing grids into smaller scales, is required for a path with the minimum cost.

6.4.2 The gridless minimum detour length router

The previous gridless routers usually provide paths with the minimum number of bends. When the minimum detour length theory is combined, a new gridless router can be proposed, that provides a path with the minimum cost in a costing workspace efficiently, because the length of an extension line, the detour length, the via cost, and the area (congestion) cost can be treated as real numbers. Considering the additional costs for area and vias, the detour length, $l'_{detour}(p_i, p_{i+1})$, is computed by:

$$l'_{detour}(p_i, p_{i+1}) = l_{detour}(p_i, p_{i+1}) + l_{cost} \tag{6.6}$$

where $l_{detour}(p_i, p_{i+1})$ is computed by *FUNCTION* l_{detour} (see Section 6.1.3), and l_{cost} is the routing length for the total additional cost:

$$l_{cost} = \frac{1}{2}[l_{areacost}(p_i, p_{i+1}) + l_{viacost}(p_i) + l_{viacost}(p_{i+1})] \qquad (6.7)$$

where $l_{areacost}(p_i, p_{i+1})$ is the additional cost for line segment $e_{p_i p_{i+1}}$, $l_{viacost}(p_i)$ is an additional cost of p_i, if p_i is a via, and $l_{viacost}(p_{i+1})$ is an additional cost of p_{i+1} if p_{i+1} is a via. If p_i and p_{i+1} are not vias, then $l_{viacost}(p_i) = l_{viacost}(p_{i+1}) = 0$. Using Equations 6.6 and 6.7, Equation 6.4 is modified as:

$$L_{detour} = \sum_{i=0}^{n-1} l'_{detour}(p_i, p_{i+1}) \qquad (6.8)$$

where L_{detour} is the cost or the routing length, $p_1 = s$ and $p_n = \tau$.

Using detour length l'_{detour}, the search process is implemented below. Let E be the set of extension lines obtained by extending all non-concave line segments of the boundary of every obstacle. The search in our router starts from s. Let $P_1 = \{p_1, p'_1, ..., p_1^{(n_1)}\}$ be a set of intersections between the source extension lines and the extension lines in E. If $l'_{detour}(s, p_1) \leq l'_{detour}(s, p_1^{(i)})$ ($i = 1, ..., n_1$), then line segment e_{sp_1} is stored as a child of s in a search tree, and the search is expanded to each extension line crossing p_1. Let $P_2 = \{p_2, p'_2, ..., p_2^{(n_2)}\}$ be a set of intersections between an extension line crossing p_1 and the extension lines in E. If $l'_{detour}(p_1, p_2) \leq l'_{detour}(p_1, p_2^{(j)})$ ($j = 1, ..., n_2$) and $l'_{detour}(s, p_1) + l'_{detour}(p_1, p_2) \leq l'_{detour}(s, p_1^{(i)})$ ($i = 1, ..., n_1$), then $e_{p_1 p_2}$ is stored as a child of p_1. The search continues until the target, τ, is found. If $e_{p_n \tau}$ is the last extension line stored for the backward search, then a path including $e_{p_n \tau}$, $e_{p_{n-1} p_n}$, ..., e_{sp_1} is a path connecting s and τ with the minimum cost.

Example: Figure 6.28 shows a gridless search application of the minimizing-detour-length theory. The workspace is the same as that shown in Figure 6.25a with $l_{cost} = 0$. The search starts from s. Because detour length from s to p_{11} or p_{12} is 0 and the detour length from s to p_{13} is greater than 0, p_{11} and p_{12} are children of s. Because $l'_{detour}(p_{11}, p_{21}) = l'_{detour}(p_{12}, p_{21}) = 0$, p_{21} is a child of p_{11} and p_{12}. Because $l'_{detour}(p_{21}, p_{22}) = 0$, p_{22} is a child of p_{21}. Because $l'_{detour}(p_{22}, p_{31}) < l'_{detour}(p_{11}, p_{13})$, p_{31} is a child of p_{22}. Because $l'_{detour}(p_{31}, p_{41}) = 0$, p_{41} is a child of p_{31}. Because $l'_{detour}(p_{41}, \tau) = 0$, the searching is successful. Two paths obtained in the backward search are shown in Figures 6.28b and 6.28c.

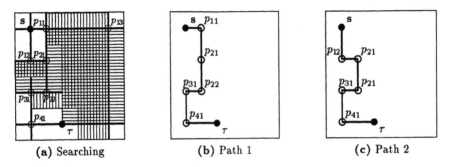

| (a) Searching | (b) Path 1 | (c) Path 2 |

Figure 6.28. A gridless minimum detour length searching

◇

6.4.3 Searching for paths with special requirements

Signal lines of high-speed circuitry must occasionally be routed with paths that satisfy the bounded length or the matched pair requirement (see Section 2.2 of Chapter 2). Signal lines must, therefore, be given a higher routing priority than other signal lines, so that a net (tree) with the special requirement can be found efficiently. In this section, two approaches using maze searching are employed to solve the bounded length problem and the matched pair problem.

6.4.3.1 The bounded-length problem

The bounded length problem for paths $e_{s_1\tau_1}$, $e_{s_2\tau_2}$ with the bounded length l_b can be expressed as $l_{e_{s_1\tau_1}} = l_{e_{s_2\tau_2}} \pm l_b$ $(l_{e_{s_2\tau_2}} < l_{e_{s_1\tau_1}})$, where s_1, s_2, τ_1 and τ_2 are terminal grids, $l_{e_{s_1\tau_1}}$ is the length of $e_{s_1\tau_1}$, and $l_{e_{s_2\tau_2}}$ is the length of $e_{s_2\tau_2}$. Let i_τ be the number labeled in the ith expansion starting from a terminal grid, $\tau \in \{s_1, s_2, \tau_1, \tau_2\}$, in the forward search. The maze search starts from s_1, s_2, τ_1 and τ_2, simultaneously, until a grid, g_1, labeled i_{s_1} and i_{τ_1}, and another grid, g_2, labeled i_{s_2} and i_{τ_2}, are found, where $i_{s_1} + i_{\tau_1} = i_{s_2} + i_{\tau_2} \pm l_b$. Backward searching from grid g_1 to s_1 and to τ_1, and from g_2 to s_2 and to τ_2, four subpaths, $e_{g_1s_1}$, $e_{g_1\tau_1}$, $e_{g_2s_2}$ and $e_{g_2\tau_2}$, are obtained. The path connecting s_1 and τ_1 with the length, $l_{e_{s_1\tau_1}} = i_{s_1} + i_{\tau_1}$, and the path connecting s_2 and τ_2 with the length, $l_{e_{s_2\tau_2}} = i_{s_2} + i_{\tau_2}$, are given as $e_{s_1\tau_1} = e_{g_1s_1} \cup e_{g_1\tau_1}$ and $e_{s_2\tau_2} = e_{g_2s_2} \cup e_{g_2\tau_2}$, respectively. If there are n_1 grids labeled i_{s_1} and i_{τ_1}, and n_2 grids labeled i_{s_2} and i_{τ_2}, where $i_{s_1} + i_{\tau_1} = i_{s_2} + i_{\tau_2} \pm l_b$, then there exists at least one and at most

$n_1 + n_2$ different pairs of paths, that satisfy two requirements: $l_{e_{s_1 \tau_1}} + l_{e_{s_2 \tau_2}}$ is the minimum, and $l_{e_{s_1 \tau_1}} = l_{e_{s_2 \tau_2}} \pm l_b$.

Example: Two cases of finding paths for the bounded length problem with $l_b = 0$ and $l_b = 2$ are shown in Figure 6.29, in which $s_1 = s_2 = s$. In the case with $l_b = 0$, g_1 is a grid labeled $\{4_s, 3_{\tau_1}\}$ or $\{3_s, 4_{\tau_1}\}$, and g_2 is a grid labeled $\{4_s, 3_{\tau_2}\}$ or $\{3_s, 4_{\tau_2}\}$ (see Figure 6.29a). Paths $e_{s\tau_1}$ and $e_{s\tau_2}$ in Figure 6.29a provide one of the solutions satisfying $l_{s\tau_1} = l_{s\tau_2}$, with the minimum length, $l_{s\tau_1} + l_{s\tau_2}$. In the case of $l_b = 2$, g_1 is a grid labeled $\{3_s, 6_{\tau_1}\}$, or $\{6_s, 3_{\tau_1}\}$, and g_2 is a grid labeled $\{a_s, b_{\tau_2}\}$ ($2 \leq a \leq 6, b = 7 - a$). Figure 6.29b shows paths connecting s and τ_1 with the minimum length longer than that of the path connecting s and τ_2 by 2. Thus, $l_{e_{s\tau_1}} + l_{e_{s\tau_2}} = 14, 15$ or 16. The pair of paths with a length equal to 14 is a minimum rectilinear Steiner tree, that satisfies the requirement of $l_b = 2$ on the the workspace in Figure 6.29.

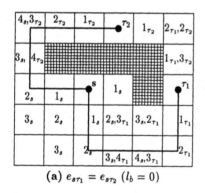
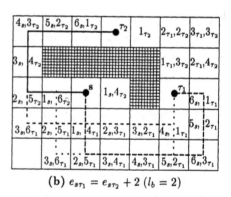

(a) $e_{s\tau_1} = e_{s\tau_2}$ ($l_b = 0$)   (b) $e_{s\tau_1} = e_{s\tau_2} + 2$ ($l_b = 2$)

Figure 6.29. Routing for the bounded length problem

◇

6.4.3.2 The matched-pair problem

The matched pair problem can be converted to a problem of searching for a pair of paths, $e_{s_1 \tau_1}$ and $e_{s_2 \tau_2}$, that satisfy the following matching condition:

If p_1 is a grid in path $e_{s_1 \tau_1}$, then there exists a grid p_2 in path $e_{s_2 \tau_2}$, that satisfies $|x_{p_1} - x_{n_1}| = |x_{p_2} - x_{n_2}|$ and $|y_{p_1} - y_{n_1}| = |y_{p_2} - y_{n_2}|$, where $n_1 \in \{s_1, \tau_1\}$ and $n_2 \in \{s_2, \tau_2\}$.

A pair of grids in two paths that satisfies the matching condition is called a pair of matched grids. If obstacles are ignored in the workspace, a path $e'_{s\tau}$ that matches another path $e_{s\tau}$ exists in an area enclosed by the rectangle with s and τ as vertices. From the matching condition, s_1 and τ_1 must match s_2 and τ_2, respectively, if edges $e_{s_1\tau_1}$ and $e_{s_2\tau_2}$ are a matched pair. That is, $|x_{s_1} - x_{\tau_1}| = |x_{s_2} - s_{\tau_2}|$ and $|y_{s_1} - y_{\tau_1}| = |y_{s_2} - y_{\tau_2}|$.

Initially, all paths for $e_{s_1\tau_1}$ and $e_{s_2\tau_2}$ are found by the router. If a path connecting s_1 and τ_1 matches a path connecting s_2 and τ_2, then a matched pair $\{e_{s_1\tau_1}, e_{s_2\tau_2}\}$ is found. If no path connecting s_1 and τ_1 matches a path connecting s_2 and τ_2, and the length of the path connecting s_1 and τ_1 is less than or equal to the length of the path connecting s_2 and τ_2, then all paths connecting s_1 and τ_1 are removed, and then replaced with a path, $e'_{s_1\tau_1}$, that matches $e_{s_2\tau_2}$. If no obstacle is crossed by path $e'_{s_1\tau_1}$, then $\{e'_{s_1\tau_1}, e_{s_2\tau_2}\}$ is a matched pair. Otherwise, available grids, crossed by path $e_{s_2\tau_2}$ and matched with the obstacles crossed by path $e'_{s_1\tau_1}$, are considered imaginary obstacles. Simultaneously expanding the obstacles crossed by $e'_{s_2\tau_2}$ and the imaginary obstacles crossed by $e'_{s_1\tau_1}$, a new matched pair with or without crossing obstacles can be found. If obstacles are crossed by a new matched pair, then imaginary obstacles are set, and the process is repeated, until either a new matched pair that does not cross any obstacles is found, or no grid is available for the expansion. If a matched pair cannot be found in the given workspace, the number of layers must be increased, so that a matched pair can be accommodated.

Example: Figure 6.30 depicts an iterative process for finding a matched pair connecting terminal pair τ_1 and s and terminal pair connecting τ_2 and s. In Figure 6.30a, $\{e_{s\tau_1}, e_{s\tau_2}\}$ is a pair of paths with the minimum length, but is not a matched pair. Because $l_{e_{s\tau_1}} < l_{e_{s\tau_2}}$, paths connecting s and τ_1 are replaced with path $e'_{s\tau_1}$, that matches path $e_{s\tau_2}$ (Figure 6.30b). Because an obstacle is crossed by $e'_{s\tau_1}$, $\{e_{s\tau_2}, e'_{s\tau_1}\}$ is not a matched pair. Two grids crossed by $e_{s\tau_2}$ match the obstacle crossed by $e'_{s\tau_1}$. Two imaginary obstacles are assigned to these grids, that are shown in newly darkening grids in Figures 6.30c and 6.30d, respectively. Then, the obstacle crossed by $e'_{s\tau_1}$ and the imaginary matched obstacle crossed by $e_{s\tau_2}$ are simultaneously expanded. From the new expansions shown in Figures 6.30c and 6.30d, two paths connecting s and τ_2 match the path connecting s and

τ_1, respectively. The two matched pairs are shown in Figures 6.30e and 6.30f, respectively. The matched pairs shown in Figure 6.30e are minimum rectilinear Steiner trees on a workspace with obstacles, with its edges satisfying the matching condition.

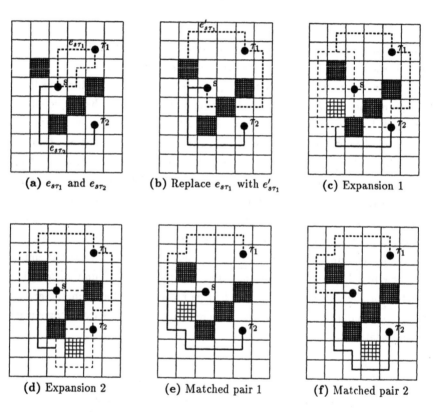

Figure 6.30. Routing for the matched pair problem

■ An original obstacle   ▦ An imaginary obstacle

◇

6.5   References

[AKE67]   Akers, S. B., "A Modification of Lee's Path Connection Algorithms," *IEEE Trans. on Electronic Computers (short notes)*, vol. EC-16, 1967, pp. 97-98.

[ASA85]   Asano, T., Sato, M. and Ohtsuki, T, "Computational Geometry Algorithms," *Layout Design and Verification*, edited by T. Ohtsuki, North-Holland, Amsterdam, 1985.

[BUR83]   Burstein, M. and Pelavin, R., "Hierarchical Wire Routing," *IEEE Trans. on Computer Aided Design*, vol. CAD-2, no. 4, Oct., 1983, pp. 223-234.

[CLO84]   Clow, G. W., "A Global Routing Algorithm for General Cells," , *Proc. of 21st Design Automation Conf.*, 1984, pp 45-51.

[DEE82]   Dees, W. A. and Karger, P. G., "Automated Rip-up and Reroute Techniques," *Proc. of 19th Design Automation Conf.*, 1982, pp 432-439.

[FIN85]   Finch, A. C., Machenzie, K. J., Balsdon, G. J. and Symonds, G., "A Method for Gridless Routing of Printed Circuit Boards," *Proc. of 22th Design Automation Conf.*, 1985, pp. 509-515.

[GEY71]   Geyer, G. M., "Connection Routing Algorithm for Printed Circuit Board," *IEEE Trans. on Circuit Theory*, vol. CT-18, 1971, pp. 95-100.

[HAD77]   Hadlock, F. O, "A Shortest Path Algorithm for Grid Graphs," *Networks*, vol. 7, 1977, pp. 323-334.

[HAN90]   Hanafusa, A., Yanashita, Y. and Yasuda, M., "Three-Dimensional Routing for Multilayer Ceramic Printed Circuit Boards," *Proc. IEEE Int. Conf. on CAD*, Nov. 1990, pp. 386-389.

[HEI68]   Heiss, S., "A Path Connection Algorithm for Multi-layer Board," *Proc. of 5th Design Automation Conf.*, 1968, pp. 1-14.

[HEY80]   Heyns, W., Sansen, W. and Beke, H., "A Line-Expansion Algorithm for the General Routing Problem with a Guaranteed Solution," *Proc. of 17th Design Automation Conf.*, 1980, pp. 243-249.

[HIG69]   Hightower, D. W., "A Solution to the Line-Routing Problem on the Continuous Plane," *Proc. of 6th Design Automation Conf.*, 1969, pp. 1-24.

[HIG74]   Hightower, D. W, "The Interconnection Problem: A Tutorial," *IEEE Comput.*, vol. C-7, no. 4, 1974, pp. 18-32.

[HOE76]   Hoel, J. H, "Some Variations of Lee's Algorithm," *IEEE Comput.*, vol. C-25, 1976, pp. 19-24.

[HSU83]   Hsu, C. P., "A New Two-Dimensional Routing Algorithm," *Proc. of 19th Design Automation Conf.*, 1983, pp. 46-50.

[HU85]   Hu, T. C., and Kuh, E. S., *VLSI Circuit Layout: Theory and Design*,
           IEEE Press, New York, 1985.

[IOS86]   Iosupovici, A., "A Class of Array Architectures for Hardware Grid
           Router," *IEEE Trans. on Computer Aided Design*, vol. CAD-5, no.
           2, 1986, pp. 245-255.

[LAU80]   Lauther, U., "A Data Structure for Gridless Routing," *Proc. of 17th
           Design Automation Conf.*, 1980, pp. 603-609.

[LEE61]   Lee, C., "An Algorithm for Path Connection and its Application," *IRE
           Trans. Electron. Comput.*, vol. EC-10(3), 1961, pp. 346-365.

[LUN88]   Lunow, R. E., "A Channelless, Multilayer Router," *Proc. 25th Design
           Automation Conf.*, 1988, pp. 667-671.

[MAR81]   Marek-Sadowska, M and Kuh, E. S., "A New Approach to Routing
           Two-layer Printed Circuit Boards," *Int. J. Circuit Theory and Appl.*,
           vol. 9, no. 3, July 1981, pp. 131-132.

[MIK68]   Mikami, K. and Tabuchi, K, "A Computer Program for Optimal Rout-
           ing of Printed Circuit Connectors," *IFIPS Proc.*, 1968, pp. 1475-1478.

[MOO59]   Moore, E. F., "The Shortest Path Through a Maze," *Proc. Int. Symp.
           Theory of Switching, Part II*, 1959, 285-290.

[OHT84]   Ohtsuki, T., and Sato, M, "Gridless Routers for Two-Layer Intercon-
           nection," *Proc. IEEE Int. Conf. on CAD*, Nov. 1984, pp. 76-78.

[OHT85]   Ohtsuki, T., "Gridless Routers – New Wire Routing Algorithms Based
           on Computational Geometry, " *Int. Conf. on Circuit and System*,
           China, 1985.

[OHT86]   Ohtsuki, T., *Layout Design and Verification*, North-Holland, Amster-
           dam, 1985.

[RUB74]   Rubin, F., "The Lee Path Connection Algorithm," *IEEE Trans on
           Comput.*, vol. C-23, 1974, pp. 907-914.

[SCH90]   Schiele, W. L., Krüger, Th., Just, K. M., and Kirsch, F. H., "A Grid-
           less Router for Industrial Design Rules," *Proc. of Design Automation
           Conf.*, 1990, pp. 626-631.

[SET90]   Seto, M., Kubota, K., and Ohtsuki, T., "A Hardware Implementation
           of Gridless Routing Based on Content Addressable Memory," *Proc. of
           Design Automation Conf.*, 1990, pp. 646-649.

[SO74]   So, H. C. "Some Theoretical Results on the Routing of Multilayer Printed Wiring Boards," *Proc. IEEE Int. Symp. Circuit Syst.*, 1974, pp. 296-303.

[SOU78]   Soukup, J., "Fast Maze Router," *Proc. of Design Automation Conf.*, 1978, pp. 100-102.

[TAD80]   Tada, T., Yoshimura, K., Kagata, T. and Shirakawa, T., " A Fast Maze Router with Iterative Use of Variable Search Space Restriction," *Proc. of Design Automation Conf.*, 1980, pp. 250-254.

[TAD86]   Tada, T. and Hanafusa, A., "Router System for Printed Wiring Boards of Very High-Speed, Very Large-Scale Computer," *Proc. of Design Automation Conf.*, 1986, pp. 791-797.

[TIN79]   Ting, B. S., Kuh E. S. and Sangiovanni-Vincentelli, A., "Via Assignment Problem in Multilayer Printed Circuit Board," *IEEE Trans. on Circuit and System*, vol. CAS-26, no. 4, 1979, pp. 261-272.

[TOM80]   Tompa, M, "An Optimal Solution to a Wire-routing Problem," *Proc. of 12th ACM Symp. on Theory of Comput.*, 1980, pp. 161-176.

[YAN72]   Yang, Y. Y. and Wing, O., "Suboptimal Algorithm for a Wire Routing Problem," *IEEE Trans. on Computer Aided Design*, vol. 1, no. 1, 1982, pp. 25-35.

Chapter 7

Via Minimization

Guoqing Li, Michael Pecht and Yeun Tsun Wong

During the physical design of integrated circuits, multi-chip module substrates, and printed circuit and wiring boards, interconnections are generally formed by patterning conductive paths on two or more layers. Because the layers are usually separated by a thin insulator, holes or vias are formed to connect the conductive paths on different layers.

Most existing routing algorithms produce a large number of vias because they solve the layer assignment problem by placing all horizontal line segments on one layer and all vertical line segments on the other. However, the number of vias should be kept small. The reduced routing real estates that result (a via typically has a diameter two to five times larger than the metallization width) increase both the ease of manufacture (vias must be mechanically or laser drilled), and performance, especially for high-speed circuits for which direction changes are undesirable.

Minimizing the number of vias generally involves finding the layer assignment that requires the fewest possible vias. This chapter discusses the via minimization problem and presents algorithms to reduce vias.

7.1   Introduction

A *via* is a particular junction where two or more layers are electrically interconnected. A valid layer assignment is a set of wire segments and vias inside the routing region that implements the definitions of the set of signal nets without violating the design rules.

In general, the via minimization problem can be considered as two categories: the *Unconstrained Via Minimization* (UVM) problem and the *Constrained Via Minimization* (CVM) problem. In the unconstrained case, via minimization, one of the optimal criteria, is accomplished during the routing steps. The UVM optimization problem can be defined as follows. Given a routing problem, $P = (M, T, N, l)$, where M, T, N are the sets of *modules, terminals,* and *nets,* respectively, and l is the number of layers, find a layout that connects all nets and uses the minimum number of vias for l layers assignment.

This approach results in the fewest possible vias. However, it is hard to implement, not only because it is a computationally difficult problem to be solved but also because the optimum solution is usually an undesirable layout with long wire lengths and a large routing area.

If, on the other hand, the topology of the routing is fixed, then the problem of finding a solution to minimize the number of vias is called a constrained via minimization (CVM) problem. In this approach, via minimization is accomplished after routing has been completed and without altering the layout generated by the previous router. Vias are set as low as possible. The CVM optimization problem can be expressed as follows. Given a layout, $L = (M, T, W, l)$, where M, T, W are the sets of *modules, terminals,* and *wire segments,* respectively, and l is the number of layers, find a layer assignment for W using l layers and a minimum subset of the possible vias that allow segments to be connected without violating design rules.

Example: Two solutions of layer assignment for a given layout are shown in Figure 7.1, where eight vias are required in (a) and two vias are required in (b), respectively. Blank circles represent vias. Dashed lines are on one layer, and solid lines are on another.

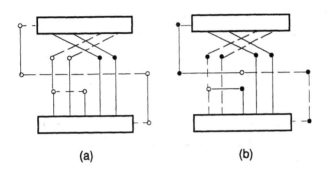

(a)   (b)

Figure 7.1. Two solutions for a layout

◇

7.2   The *NP*-Completeness of the CVM Problem

NP-complete to the CVM problem means that the CVM problem is as hard as all the problems in the class of the *NP*-complete problem, and that it has no polynomial time solution. The CVM problem remains *NP*-complete even when some restrictions such as grid-based layout, via location, and junction degree, are added.

This section deals with the complexity of the CVM problem and shows that the associated decision problem is *NP*-complete in general case [NAC87A].

7.2.1   The general 2-CVM problem

Given a layout $L = (M, T, W)$ and a positive integer K, where M, T, W are sets of *modules, terminals,* and *wire segments* respectively, the 2-CVM problem asks whether a valid layer assignment exists for W using two layers such that there are at most K vias. The 2-CVM is in $\mathcal{NP}$ because K may be bounded by n^2, where n is the number of routing segments. Thus, a nondeterministic algorithm need only guess a solution with K or fewer vias and then check for validity in polynomial time. For a problem to be NP-complete, a known NP-complete problem must be transformable to it. The known problem chosen here is the *planar node cove* (PNC) problem. That is, for a given planar graph $G = (N, E)$ and an positive integer K, is there a subset $N' \subseteq N$ such that for all edges $(u, v) \in E$, either $u \in N'$ or $v \in N'$, and $|N'| \leq K$?

The PNC was shown to be NP-complete by Garey and Johnson [GAR79]. The proof of PNC $\propto$ 2-CVM will essentially show how to start from an arbitrary instance I of PNC and construct, in polynomial time, an instance I' of 2-CVM such that K or fewer is satisfied in PNC if and only if the instance I' can be generated with K' or fewer vias in 2-CVM.

To construct the transformation, a sublayout B is defined, as shown in Figure 7.2, where s and t are junctions. Sublayout B requires at least one via if it is to be realized in two layers. If a via is placed at either s or t, then B will have a valid layer assignment.

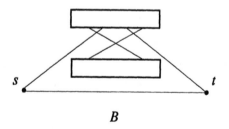

$$B$$

Figure 7.2. Sublayout B

Lemma 7.1. Given a layout L that consists of one or more nonoverlapping sublayouts joined at either junction s or junction t, then there will be a valid layer assignment using two layers for L if either s or t is replaced in each sublayout with a via.

Proof: Because layout L consists of nonoverlapping sublayouts, each sublayout can be considered an independent sublayout. Routing segments in different sublayouts intersect only at vias, either s or t. Therefore, routing segments within a sublayout can be assigned to a layer without considering the layer assignment of segments that are not in this sublayout. Because a valid assignment exists for a sublayout providing the via at s or t, and because all sublayouts in L are not overlapping, the entire layout L has a valid layer assignment. $\square$

Example: In Figure 7.3, small circles represent vias, solid lines represent the segments on one layer, and dashed lines represent the segments on another layer.

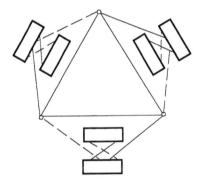

Figure 7.3. An example of Lemma 7.1

$\diamond$

The following theorem can be proved by constructing a transformation, which can be performed in polynomial time, from an instance in PNC to an instance in 2-CVM.

Theorem 7.1. 2-CVM is NP-complete.

Proof: Suppose there is an instance $G = (N, E)$ and K of PNC. Without loss of generality, graph G may be assumed to be connected. The instance constructed for the 2-CVM problem is a layout, $L(G)$, with the same integer, K. Let G' be a straight-line planar embedding of the graph G. Such embedding can be obtained from any planar graph in polynomial time [FAR48]. The line segments and points in G' are referred to by corresponding edges and nodes in G. Because no two edges intersect in G', a small region can be formed around each edge $(s', t') \in G'$ such that no two regions overlap. Then layout $L(G)$ can be obtained by replacing each edge (s', t') with a sublayout B, such that B is completely inside the region, and s and t in B correspond to s' and t' in edge (s', t'), respectively.

Assume that there is a node cover, $N' \subseteq N$, such that $|N'| \leq K$. N' must include at least one node for each edge $(s', t') \in E$ by definition. The set of vias, V, can be constructed as follows. If the corresponding node for each junction is in N', a via is placed. Thus, either s or t (or both) will be included for each sublayout B in V and $|N'| = |V|$. Therefore, there is a set of vias of size K or less such that a valid layer assignment exists for $L(G)$ by Lemma 7.1.

Conversely, suppose there is a desired $L(G)$ for which a valid layer assignment exists with a set of vias V, $|V| \leq K$. The preceding construction and Lemma 7.1 indicate that V must include at least one via for each sublayout. All vias can be assumed to be located at the junction s or t because they can be moved in the same sublayout without increasing the vias or affecting the valid layer assignment. Because the sublayouts only intersect at s or t, the via can reasonably be put at s or t. Let N' be the set of nodes that have corresponding vias in V. Then for each edge (s', t'), either $s' \in N'$, or $t' \in N'$ (or both). Therefore, a node cover, N', for G is generated with $|N'| = |V| \leq K$.

Thus, the proof of the theorem is complete because the transformation from PNC to 2-CVM can be performed in polynomial time. The only requirement is replacing each edge in the straight-line planar embedding of G with the sublayout B. $\square$

Example: An instance of PNC with $K = 2$ is shown in Figure 7.4a, where $N' = \{n_1, n_2\}$. Figure 7.4b shows the corresponding layout with $V = \{v_1, v_2\}$.

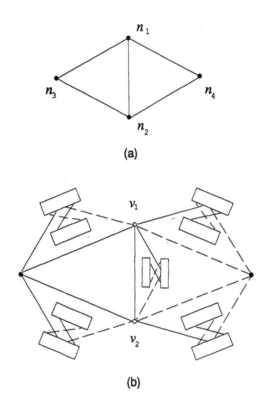

(a)

(b)

Figure 7.4. Construction for Theorem 7.1

◇

7.2.2   Restricted 2-CVM problem

The 2-CVM problem in restricted cases is still NP-complete. The problem was originally defined in [CHE83]. Further restriction of the maximum junction degree limit of four is specified in [CHO89]. This problem is called the 2-CVM4 problem and can be defined as follows.

For a given layout, $L = (M, T, W)$, with maximum junction degree four and a positive integer K, where M, T and W are sets of *modules, terminals,* and *wire segments*, respectively, does there exist a valid layer assignment for W using two layers such that it requires at most K vias? The proof of its NP-completeness is based on a transformation of the known NP-complete VDB4 problem for a planar graph. Given a graph $G = (N, E)$ and an integer $K \geq 0$, the *vertex-deletion graph bipartization* (VDB) problem [KRI79] involves finding a set of K or fewer vertices, $N' \subset N$, in G such that the subgraph $G(N - N')$ is bipartite, or equivalently, free of odd-length cycles [EVE79]. The planar VDB4 problem is the problem of VDB for a planar graph with a maximum degree of vertex limited to four.

The VDB4 problem can be stated as follows. Given a planar graph, $G = (N, E)$, with a maximum vertex degree of four and a positive integer K, does there exist a subset $N' \subseteq N$ such that the subgraph $G(N - N')$ is bipartite and $|N'| \leq K$? To construct the transformation from VDB4 to 2-CVM4, a sublayout H is defined as shown in Figure 7.5. It consists of a single module, m, with two terminals, t_1 and t_2, on its boundary. Wire segments w_1 and w_2 connect terminals t_1 and t_2 to junctions j_1 and j_2, respectively.

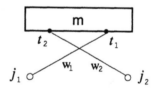

Figure 7.5. Sublayout H

For a planar graph $G = (N, E)$, a straight-line embedding of G can be obtained in which every edge in G is represented by a straight-line segment. Then the layout $L(G)$ can be constructed by replacing each edge (or straight-line segment) with a sublayout H such that junctions j_1 and j_2 coincide with two endpoints of the corresponding edge.

Example: Figure 7.6a shows a graph G. Figure 7.6b shows the layout $L(G)$ constructed from G.

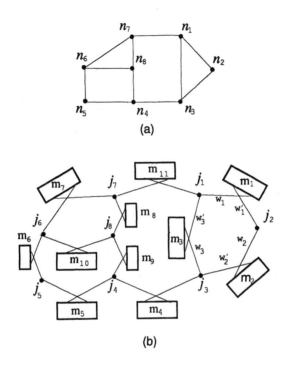

Figure 7.6. Constructing layout

◇

Definition 7.1. For each fundamental cycle C in graph G, a corresponding cycle, H-cycle, consists of sublayout H's.

Definition 7.2. An H-cycle is even (odd) if the number of H's in the H-cycle is even (odd).

Definition 7.3. A layer assignment of a wire segment w is described by a mapping, $l : w \rightarrow \{0, 1\}$, where w is embedded on layer $i, i = 1, 2$.

Lemma 7.2. A valid layer assignment for layout $L(G)$ exists without any via if and only if no odd H-cycle occurs in $L(G)$.

Proof: Suppose there is a valid layout assignment without any via. Consider the H-cycle consisting of modules m_1, m_2 and m_3, as shown in Figure 7.6. Assume

$l(w_1) = 1$, then $l(w_1') = 2$. To avoid a via at junction $j_2, l(w_2)$ should be 2. Consequently, $l(w_2') = 1, l(w_3) = 1$, and $l(w_3') = 2$. In order to electrically connect w_1 and w_3', a via must be placed at junction j_1. Thus, a via is always needed if there is an odd H-cycle in $L(G)$.

Conversely, suppose there is no odd H-cycle in $L(G)$. Begin by choosing an arbitrary segment and all of its adjacent segments. Let W_i be such a set of segments. Assign all $w_i \in W_i$ to layer 1. Find $W_i' = \{w_i'|w_i'$ is not assigned and crosses $w_i \in W_i\}$ and the set W_j containing all the segments in W_i' and their adjacent segments. Assign all $w_j \in W_j$ to layer 2. Then find $W_j' = \{w_j'|w_j'$ is not assigned and crosses $w_j \in W_j\}$ and the set W_k containing all the segments in W_j' and their adjacent segments. Assign all $w_k \in W_k$ to layer 1. Repeat this procedure until all segments are assigned. There is no conflict with layer assignments because all H-cycles are even. □

By referring to Lemma 7.2, the following theorem can be proved. Because the 2-CVM4 problem is in $\mathcal{NP}$, so it remains only to show that there is a polynomial time transformation from VDB4 to 2-CVM4.

Theorem 7.2. 2-CVM4 is NP-complete.

Proof: Let $G = (N, E)$ be a planar graph with a maximum vertex degree of four, and K be a nonnegative integer. A layout $L(G)$ is constructed as described above. Suppose that $N' \subset N, |N'| \leq K$, and $G(N - N')$ is bipartite. Let $J = \{j_i|j_i$ is a junction corresponding to $n_i \in N'\}$. The first step in transformation is to delete all $j_i \in J$ and all wire segments incident upon them from $L(G)$. $L'(G)$ denotes the resulting layout. There are two types of "degenerated" sublayouts in $L'(G)$: sublayout H_{d1} consists of a module and one wire segment; sublayout H_{d2} consists no wire segment. The next step is to delete all H_{d1} and H_{d2} from $L'(G)$. Let $L''(G)$ be the resulting layout, and G'' be the degenerated graph after all $n \in N'$ and edges incident upon them are deleted. Because a one-to-one correspondence exists between the edges in G'' and the sublayouts Hs in $L''(G)$, and because G'' is free of odd-length cycles, $L''(G)$ contains no odd H-cycle. By Lemma 7.2, a valid layout assignment exists for $L''(G)$ without any via. Then a valid layer assignment for $L'(G)$ can be obtained by assigning the only wire of each H_{d1} to the same layer as that of all of its adjacent wire segments in $L''(G)$. Now a valid layer assignment for $L(G)$ can be obtained, based on the layer assignment for $L'(G)$. A via is placed for each $j_i \in J$. Consider the wire segments incident upon the $j_i \in J$. For each wire segment that is missing from H_{d1}, if the remaining wire segment is assigned to layer 1, the missing segment is assigned to layer 2 and vice versa. For H_{d2}, two wire segments can be assigned to different layers arbitrarily. Therefore, a valid layer assignment exists for $L(G)$ that requires K or fewer vias because $|J| \leq K$.

Conversely, suppose that there is a set of K or fewer vias for which a valid layer assignment exists for $L(G)$. Let $J = \{j_i | j_i$ is a junction where a via is placed$\}$ and $N' = \{n_i | n_i \in G$ corresponding to $j_i \in J\}$. The layout $L(G(N - N'))$ is obtained by deleting all H's which contain at least one wire segment incident upon some $j_i \in J$. Note that there is not any via in $L(G(N - N'))$. Because the layer assignment for $L(G(N - N'))$ is valid and requires no via, no odd H-cycle is contained in $L(G(N - N'))$, according to Lemma 7.2. No odd-length cycle exists in the corresponding graph $G(N - N')$; hence, $G(N - N')$ is bipartite. Because $|N'| = |J|$, a set of vertices N' exists such that $G(N - N')$ is bipartite with $|N'| \leq K$.

Because constructing $L(G)$ (by replacing each segment in a straight-line planar embedding of G with a sublayout H) and obtaining the straight-line planar embedding of G can be both accomplished in polynomial time [FAR48, CHI85], the total transformation can be performed in polynomial time. Furthermore, the junction degree is the same as the degree of the corresponding vertex in the planar embedding of graph G. Because the maximum vertex degree is four in G, the maximum junction degree in $L(G)$ is also four. Thus, the proof of the theorem is complete. $\square$

Example: The construction for Theorem 7.2 is shown in Figure 7.7, where $J = \{j_1, j_7\}$, corresponding to $N' = \{n_1, n_7\}$ in graph G (Figure 7.6a). Layout $L'(G)$ and its layer assignment are shown in (a) and (c), respectively. Layer

assignment for $L''(G)$ is shown in (b). Layer assignment for $L(G)$ is shown in (d).
Sublayouts containing m_1, m_3, m_7, and m_8 in (a) are type 1 degenerated sublayouts, H_{d1}. Sublayout containing m_{11} is type two degenerated sublayout, H_{d2}.

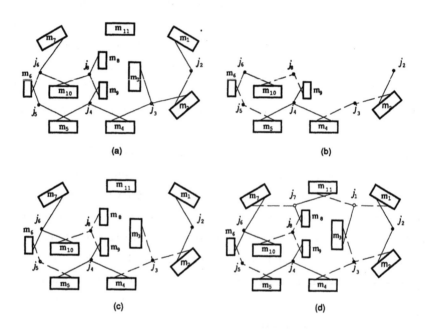

Figure 7.7. An example of Theorem 7.2

◊

Using the same argument as in [NAC87A], the following theorem holds for the
CVM problem with two layers.

Theorem 7.3. For one or more of the following restrictions, 2-CVM with a
maximum junction degree limited to four or more is still NP-complete if

1. the input layout is grid based;

2. vias are located only at the junctions that existed in the input layout.

7.2.3   The n-CVM problem

Most research and literature study the special case when there are only two layers available to embed signal nets (2-CVM problem). This section considers the via minimization problem for a layer number greater than two [MOL87].

The n-CVM problem can be stated as follows. Given a circuit in the Euclidian plane, how can the segments of the signal nets be assigned in the layers so that the total number of layer changes (i.e., the vias required) is minimized?

The signal nets of a circuit can be represented by a planar undirected graph, $G = (N, E)$, which is always possible if the crossovers are considered as vertices of the graph. For a given vertex $v \in N$, three definitions are given:

Definition 7.4. $E_v = \{(v,w)|(v,w) \in E$ for all $w \in N\}$. E_v is the set of edges with one endpoint, v.

Definition 7.5. $S_v : E_v \rightarrow \{1, 2, \cdots, n\}$ is a mapping expressed by $S_v(v,w) = i$, where the edge (v,w) is assigned to the layer $i \in \{1, 2, \cdots, n\}$.

Definition 7.6. For any vertex $v \in N$, and $(v,u), (v,t) \in E_v$, there exists a relation $(v,u) \sim_v (v,t)$ if and only if (v,u) and (v,t) are assigned to the same layer.

A n-layer assignment of G corresponding to $\{\sim_v |v \in N\}$ is described by a set, $S = \{S_v|v \in N\}$, and called a $(G, \sim)$-n-layer-assignment. Some edge in E may be embedded into different layers. Molitor calls such an edge s-*critical* [MOL87]. Obviously, a contact or via is needed in this case. *Zero-contact* or *zero-via* means the number of s-*critical* edges in G is zero and no contact (via) is required. The zero-via (zero-contact) problem for n layers is finding a $(G, \sim)$-n-layer-assignment, S, such that the number of vias (contacts) required is zero. There is a corresponding decision problem: Given an undirected planar graph $G = (N, E)$ and the relation $\sim$, does there exist a $(G, \sim)$-n-layer-assignment S such that no via is required?

Before proceeding to the proof of the NP completeness of the n-CVM problem, the zero-via problem for three layers will be considered. Because the problem is in $\mathcal{NP}$, it only needs to transform a known NP-complete problem in polynomial time to complete the proof of the NP-completeness. The chosen problem is the *planar graph 3-colorability* problem. For a given planar graph, $G(N, E)$, and three colors, 3-Colorability is to assign each vertex a color such that adjacent nodes have different colors. the planar graph 3-colorability problem asks: Given a planar graph $G = (N, E)$, is G 3-colorable? That is, does there exist a function $f : N \rightarrow \{1, 2, 3\}$ such that $f(u) \neq f(v)$ whenever $(u, v) \in E$? The problem has already been proven NP-complete in [GAR79].

The following lemma is proved by constructing a transformation from an instance of the zero-via problem to an instance of the planar graph 3-colorability problem. The transformation can be performed in polynomial time.

Lemma 7.3. The zero-via problem for three layers is NP-complete.

Proof: Let $G = (N, E)$ be a planar graph. Construct a graph, $G' = (N', E')$, by inserting a vertex on each edge in G as follows:

$$N' = N \cup N_I$$
$$N_I = \{v_{ij}| \text{ there is an edge } (v_i, v_j) \in E\}$$
$$E' = \{(v_i, v_{ij}), (v_j, v_{ij})|(v_i, v_j) \in E \text{ and } v_{ij} \in N_I\}$$

The relation between edges exists. For all $v \in N', (u, v)$ and (w, v) are assigned the same layer for all u and w if and only if $v \in N$.

Suppose for graph G there is a function $f : N \rightarrow \{1, 2, 3\}$ such that $f(u) \neq f(v)$ whenever $(u, v) \in E$. In the graph G', if $v \in N$, all the edges (v, x) incident upon the vertex v are assigned the layer with the number $f(v)$. That is, S is defined by $S_v(v, w) = f(v)$ for all $(v, w) \in E'_v$. Thus, S is a $(G', \sim)$-3-layer-assignment without an *s-critical* edge. In the sense that it can be assigned in three layers without any via.

Conversely, suppose there is a desired layer assignment for G' without any via. All edges in G' are assigned a unique layer. Because of the relation between edges described above, for any $v \in N$, all (v, x) are assigned on the same layer, where

$x \in V_I$. Thus, function $f : N \rightarrow \{0, 1, 2\}$ can be defined by $f(v) = S_v(v, x)$, where (v, x) is some edge incident upon v and $v \in N$. It follows that $f(u) \neq f(v)$ whenever $(u, v) \in E$. Thus, the zero-via problem for three layers is NP-complete.
$\square$

According to Lemma 7.3, Theorem 7.4 is valid.

Theorem 7.4. 3-CVM is NP-complete.

The rest of this section shows the n-layer zero-via problem for all $n \geq 3$. The proof is based on the induction on the number of layers.

Lemma 7.4. The zero-via problem for n layers is NP-complete for all $n \geq 3$.

Proof: For $n = 3$, see the proof of Lemma 7.3. For $n > 3$, let $G = (N, E)$ be an undirected planar graph and $\sim = \{\sim_v | v \in N\}$ be a set of relations indicating whether two edges adjacent to the v should be embedded on the same layer. Construct graph $G' = (N', E')$ by adding a new vertex v_{ij} for each pair of adjacent vertices, v_i and v_j, in G as follows (see Figure 7.8).

$$N' = N \cup N_A$$
$$N_A = \{v_{ij} | v_i, v_j \in N \text{ and they are adjacent}\}$$
$$E' = E \cup E_A$$
$$E_A = \{(v_i, v_{ij}), (v_j, v_{ij}) | (v_i, v_j) \in E \text{ and corresponding } v_{ij} \in N_A\}$$

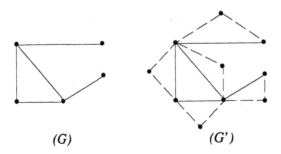

(G)   (G')

Figure 7.8. Construction of G' from G for Lemma 7.4

For all $v \in N'$, and all $(v, x), (v, y) \in E'_v$, a relation is defined as follows:

$$(v, x) \sim'_v (v, y) \ iff \ ((v, x), (v, y) \in E \ \text{and} \ (v, x) \sim_v (v, y)) \ \text{or}$$
$$(v, x), (v, y) \in E' \setminus E$$

That is, for all (v, x) and $(v, y) \in E$ with the common endpoint v, if they are assigned to the same layer in G, they are assigned to the same layer in G'. All edges in E_A are assigned to the same layer. Let $\sim' = \{\sim'_v \ |v \in N'\}$, then the following claim holds.

Claim. A $(G, \sim)$-$(n-1)$-layer-assignment S with zero via exists if and only if a $(G', \sim')$-n-layer-assignment S' exists with zero via.

Proof: Suppose a $(n-1)$-layer assignment for G exists with the set of relations $\sim$ without any via. Assign each edge $(v, x) \in E' \setminus E$ in G' the layer with the number n. The set S' is defined by $S'_v(v, w) = n$ for all $(v, w) \in E' \setminus E$, and $S'_v(v, w) = S_v(v, w)$ for all edges $(v, w) \in E$. Therefore, S' is a $(G', \sim')$-n-layer-assignment without an *s-critical* edge. That is, G' can be assigned in n layers without any via.

Conversely, suppose there is a $(G', \sim')$-n-layer-assignment S' for G' without a via. According to the definition of $\sim'$ described above, all the edges $(v, x) \in E' \setminus E$ are assigned to the same layer. Thus, the assignment of G can be obtained by deleting these edges and corresponding vertices added. In this way S can easily be obtained from S'. Therefore, a $(G, \sim)$-$(n-1)$-layer-assignment S exists without an *s-critical* edge. That is, a n-layer assignment exists without a via for the graph G. The proof of the claim is complete.

According to the claim, the zero-via problem for n layers is NP-complete for all $n \geq 3$. □

According to Lemma 7.4, the theorem below for the n-CVM problem holds, where $n \geq 3$ is the number of layers.

Theorem 7.5. n-CVM is NP-complete for all $n \geq 3$.

7.3 The NP-Completeness of the UVM Problem

This section considers the *Unconstrained Via Minimization* (UVM) problem in a two-layers routing environment. The goal is to minimize the number of vias needed to route n two-terminal nets in a bounded routing region. The associated decision problem is NP-complete [SAR89].

In a two-layer routing environment with two-terminal nets, each net of the circuit contains only two terminals. The layer assignment is obtained by mapping pieces of wires into one of the two layers. Vias are placed when a wire has to change layers to avoid crossing other wires.

Example : Two assignments in two layers corresponding to the same problem are illustrated in Figure 7.9, where (a) contains three vias and (b) contains two vias, respectively. Solid lines and dashed lines represent the curves in layer 1 and layer 2 respectively, and small circles represent the established vias.

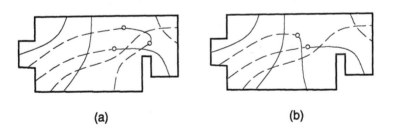

(a)   (b)

Figure 7.9. Two assignments of layers

◇

7.3.1 Basic definitions

Via minimization entails finding a routing topology of a set of n two-terminal nets in a two-layer routing region such that the necessary number of vias is minimal. The problem is called a *two-layer bounded region UVM problem*. It is equivalent to the *maximum two-independent set problem* in a circle graph, where the circle corresponds to a bounded routing region with terminals on the boundary, and each chord corresponds to a two-terminal net. The circle graph introduced in [EVE71] explicitly capturesthe wire-crossing constraint that is the key constraint to observe in solving the UVM problem. Before considering the complexity of the UVM problem, several related definitions are given:

Definition 7.7. Two chords, N_1 and N_2, on a circle are independent if they do not intersect.

Definition 7.8. In a set of chords, a subset, S_1, is independent if N_i and N_j are independent for all N_i and N_j in S_1, $i \neq j$.

Definition 7.9. A subset of chords, S_1, forms a clique if N_i and N_j intersect for all N_i and N_j in S_1, $i \neq j$.

Definition 7.10. A set of chords is two-independent if it can be partitioned into two independent subsets of chords.

An independent subset S_1 is maximal if no chord can be added into S_1 such that the resulting set is still independent. The maximum two-independent set problem in a circle graph is to find a two-independent subset of chords, S_1, such that $|S_1|$ is maximum. Obviously, two independent subsets of chords in S_1 determine the layer assignment of corresponding nets. The nets corresponding to one independent subset can be assigned in one layer, and other nets corresponding to another independent subset can be assigned another layer, so that no via is required for all these nets.

7.3.2   The NP-completeness of the two-layer UVM problem

Once the maximum two-independent set is found, the layer assignment using a minimum number of vias can be obtained, based on Lemma 7.5 [MAR84].

Lemma 7.5. An optimal solution exists for the two-layer bounded region UVM problem such that each net uses, at most, one via.

For n given chords, the two-independent set problem is to determine if there is a two-independent set of chords of size at least K. Following part shows that the problem is NP-complete. Therefore, the two-layer bounded region UVM and the general UVM problem are NP-complete. The corresponding decision problem can be stated: Given n chords in a circle and a positive integer K, does a two-independent set of chords, S_1, exists such that $|S_1| \geq K$?

The proof of NP-completeness is based on transforming a known NP-complete planar 3SAT problem (P3SAT) in polynomial time [LIC82]. The 3SAT problem can be described as follows. A *Boolean expression* is an expression composed of variables, parentheses, and the operator $\wedge$ (logical AND), $\vee$ (logical OR), and $\neg$ (negation). Variables take on value 0 (false) or 1 (true); so do expressions. An expression is *satisfiable* if there is some assignment of 0s and 1s to the variables that gives expression the value 1. The *satisfiability problem* is to determine, given a Boolean expression, whether it is satisfiable. A Boolean expression is said to be in *conjunctive normal form* (CNF) if it is of the form $E_1 \wedge E_2 \wedge \cdots \wedge E_k$, and each E_i, called a *clause*, is of the form $u_{i_1} \vee u_{i_2} \vee \cdots \vee u_{i_r}$, where each u_{i_j} is a literal, either v or $\neg v$, for some variable v. Usually $\neg x$ is written as $\bar{x}$. The expression is said to be in 3-CNF if each clause contains exactly three distinct literals. Let $C = \{c_1, c_2, \cdots, c_m\}$ in 3-CNF, where each clause is a subset of three literals from the sets $V = \{v_1, v_2, \cdots, v_n\}$ and $\overline{V} = \{\bar{v}_1, \bar{v}_2, \cdots, \bar{v}_n\}$. The 3SAT problem is to determine if a truth assignment exists for the variables such that the formula C has a truth value of 1 (true).

For a given C, graph $G_c = (N, E)$ is constructed from C as follows:

$$N = \{c_j | 1 \leq j \leq m\} \cup \{v_i | 1 \leq i \leq n\}$$
$$E = \{(c_j, v_i) | v_i \in c_j \text{ or } \bar{v}_i \in c_j\} \cup \{(v_i, v_{i+1}) | 1 \leq i \leq n, v_1 = v_{n+1}\}$$

Example: For a given $C = \{c_1, c_2\}$, G_c is constructed as demonstrated in Figure 7.10, where $c_1 = v_1 \vee v_2 \vee v_3$ and $c_2 = v_1 \vee v_3 \vee v_4$. Notice that $(v_1 \vee \bar{v}_2 \vee v_3)$ would give the same graph as $(v_1 \vee v_2 \vee v_3)$.

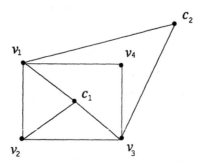

Figure 7.10. Construction of G_c from $C = \{c_1, c_2\}$

◇

P3SAT is 3SAT restricted to G_c such that G_c is planar. That is, for a given set $V = \{v_i | 1 \leq i \leq n\}$ of n boolean variables and a set $C = \{c_j | 1 \leq j \leq m\}$ in 3-CNF over V, where the graph G_c is planar, is C satisfiable? Is there a truth assignment for V such that each clause in C is true? The NP-completeness of the two-independent set problem will be proved by transforming the P3SAT problem.

Theorem 7.6. The two-independent set problem is NP-complete.

Proof: The problem is in $\mathcal{NP}$ by guessing K chords and then verifying if they are two-independent. To show its NP-completeness, a given instance of P3SAT will be transformed to an instance of the two-independent set problem such that P3SAT is satisfiable if and only if the two-independent set contains at least K chords to be determined later. Because G_c is planar, vertices corresponding to clauses may be placed inside or outside the cycle of variables. This section considers a case involving only inner vertices. For the outside case, the construction can be modified to handle outer vertices [SAR89]. The construction is as follows:

- For the cycle of variables in G_c, draw a sufficiently large circle.

- For each variable $v_i \in G_c$, draw a cycle of $2k_i$ chords $N_t^i, t = 1, 2, \cdots, 2k_i$, called *variable chords* numbered from 1 to $2k_i$ in counterclockwise order, where k_i is the degree of the vertex v_i in G_c minus 1. Even-numbered chords (solid lines) are referred to as *true chords*, corresponding to the truth value 1, and odd-numbered chords (dashed lines) are referred to as *false chords*, corresponding to the truth value 0. N_1^i is a long variable chord used only to complete the cycle of variables. Because variable chords are used in pairs in the construction, the chord N_2^i is also used as a dummy chord. All *true chords* and all *false chords* are independent. However, each *true chord* is crossed with its two neighboring *false chords*, and vice versa. A weight is defined for a chord. For example, the chord with a weight w consists of w parallel (or independent) chords. Each variable chord has a weight m, where m is the number of clauses. Weight-m chords are represented by solid and dashed lines of medium thickness, as shown in Figure 7.11.

- For each variable $v_i \in G_c$, there are $2k_i$ *superchords*, $Sup\ N_t^i, t = 1, 2, \cdots, 2k_i$. Each has a weight, $2m$, and forms a clique with two consecutive true and false chords, N_t^i and $N_{t+1}^i, t = 1, 2, \cdots, 2k_i$, where $N_{2k_i+1}^i = N_1^i$. They are drawn as thick lines in the figure.

- For each clause $c_l = x_i \lor x_j \lor x_k$, where each literal x_t is either the variable v_t or $\bar{v}_t$, three *clause chords* are constructed to form a clique as follows: If literal $x_t = v_t$, a chord is constructed to intersect a false chord of v_t, otherwise, it is constructed to intersect a true chord of v_t. The construction arbitrarily chooses literal x_j and assumes $x_j = v_j$. Consider a false chord N_t^j, where t is odd and $t > 1$. Its neighboring true chord N_{t+1}^j, and the superchord $Sup\ N_t^j$, form a clique with it. The chord corresponding to x_j is constructed to be independent of all the true chords, with one end point located between the two endpoints of $Sup\ N_t^j$, and to form a clique with N_t^j and $Sup\ N_t^j$, as shown in Figure 7.11b, where clause chords are represented by thin lines. The other endpoint is placed between $Sup\ N_t^j$ and $Sup\ N_{t+1}^j$, which is referred to as the *clause point* of x_j. For the other two literals, x_i and x_k, in the clause c_l, the two corresponding clause chords must form a clique with the clause chord corresponding to x_j. Their clause points also lie between $Sup\ N_i^j$ and $Sup\ N_{i+1}^j$. The other two endpoints of the two chords corresponding to x_i and x_k are determined as follows: If $x_i = \overline{v}_i$, then the endpoint of its chord is located between two endpoints of $Sup\ N_{2r-1}^i$, and the chord forms a clique with the true chord N_{2r}^i of v_i and $Sup\ N_{2r-1}^i$, where $2 \le r \le k_i$. If $x_i = v_i$, it can be handled in a symmetric manner. The clause chord corresponding to x_k is constructed similarly.

- For each clause, an *enforcer chord* with a weight $2m$ is constructed and located so that the clause points of each clause are included. They are represented by dashed lines of medium thickness in the figure. Each *enforcer chord* is independent of every other chord, except for the related three clause chords. Thus, each individual *enforcer chord* and the three clause chords form a clique of size 4.

Figure 7.11 demonstrates chord construction from an instance of P3SAT, where (a) shows variable chords corresponding to v_i with $k_i = 3$; and (b) shows clause chords of $c_l = x_i \lor x_j \lor x_k$ with $x_i = \overline{v}_i$ and $x_j = v_j$.

Let $K = K_1/2 + K_2 + K_3 + m$, where $K_1 = \sum_{i=1}^n 2k_i m$ is the total weight of the *variable chords* ($2k_i$ chords each has a weight m for v_i); $K_2 = \sum_{i=1}^n 2k_i 2m$ is the total weight of the *superchords* ($2k_i$ superchords each has a weight $2m$ for v_i); and $K_3 = 2mm$ (m enforcer chords each has a weight $2m$). Thus, the instance of the two-independent set problem has been constructed such that it has at least K chords if and only if the P3SAT is satisfiable. To check if this is the case, assume the given instance of P3SAT is satisfiable and fix an assignment satisfying the fomula of m clauses. Each clause must have a literal whose value is 1. By selecting the true or false chords according to the truth assignment of the variables, all the *superchords* and *enforcer chords*, pluse one chord for each clause, results in the creation of a two-independent set with K chords.

Conversely, assume there is a two-independent set with K or more chords. Because the *enforcer chords* are included, and because each one forms a clique of size four with the related clause chords, at most one clause chord may be included.

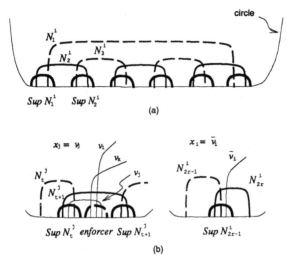

Figure 7.11. Chord construction

Therefore, the chords with weight $w > 1$ must comprise the entire solution. Otherwise, the missing chords cannot be replaced by extra clause chords, and we cannot have K nets in the solution. The variable chords must also be selected consistently. Either all true chords or all false chords of a variable should be selected in the solution, but not both. Otherwise, at least one true chord or false chord (with weight m) is lost. This chord cannot be replaced by an equal number of clause chords, because a maximum of one chord can be included in each clause. Thus, at least $K = K_1/2 + K_2 + K_3 + m$ chords exist for the structure shown in Figure 7.11b. Exactly one chord per clause is included in the solution. The chord of this clause intersects the true or false chords of the related vaiable chords in the solution. Depending on whether the long variable chord of a variable exists in the two-independent set, we can assign value 1 or 0 to the corresponding literal. For this assignment, formula C has value 1, and the P3SAT problem is satisfiable. The construction can be performed in polynomial time. Thus, the two-independent set problem is NP-complete. $\square$

Example: Theorem 7.6 is illustrated in Figure 7.12. Figure 7.12a shows the

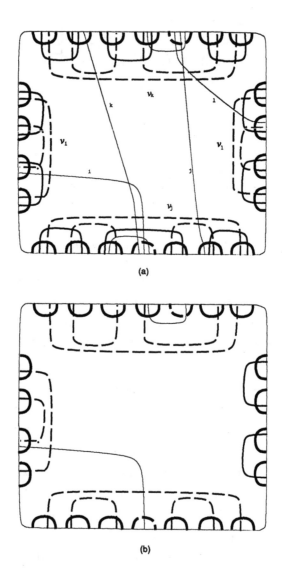

(a)

(b)

Figure 7.12. An illustration of Theorem 7.6

construction of $C = (\overline{v}_i \vee v_j \vee v_k) \wedge (v_j \vee \overline{v}_k \vee \overline{v}_l)$. Figure 7.12b shows the corresponding two-independent set, where $v_i = 0, v_j = 0, v_k = 0$, and $v_l = 1$.

◊

A layer assignment corresponding to a solution is also readily obtained. Because all clause chords in a solution are pairwise independent (due to the planarity of the instance in P3SAT), they may be assigned to layer 1. If the long variable chord N_1^i corresponding to literal v_i is in the solution, the chord only intersects a subset of clause chords and two *superchords*. These two *superchords* do not intersect any other chords because N_2^i and $N_{2k_i}^i$ are not in the solution. Thus, all long variable chords may be assigned to layer 2 if they are in the solution. The assignment of the rest of the chords is determined by the assignment of the clause chords and long variable chords.

Both CVM and UVM problems are demonstrably NP-complete and, hence, likely to be intractable. Consequently, the next step in the problems is to focus on heuristic methods of reducing the vias. The quest for the exact value of the vias will have to be restricted to promising special cases.

7.4 Algorithms and Implementations

Various algorithms implement via minimization problems, with or without restrictions [STA90, RIM87, NAC87B, CON91, DEO89]. This section introduces and discusses some of these algorithms.

7.4.1 Crossing graph and via minimization algorithms

Most existing optimal algorithms for the CVM problem with a maximum of three-way split points are based on the maximum cut algorithms for planar graphs. These algorithms have time complexities $O(m^{2.5})$, where m is the number of wire-segment clusters [CHE82, PIN82, CHE83]. A wire-segment cluster is formed by a set of wire segments such that, if one wire segment in the set is assigned a layer, the layer assignment of the rest of the wire segments in the cluster is fixed.

One such algorithm is presented by Du [DU84]. The algorithm has a time complexity $O(pn^2)$, where p is the number of vias selected, and n is the number of wire segments formed in the final layout. In the worst case, p can be $O(n^2)$. Because these algorithms require a long time to generate results for a large number of nets, more efficient and practical heuristic algorithms must be developed [CHA87].

Foundations and Crossing Graphs

A "possible via" denotes the location in a given layout that is the possible position for a via, as specified by the user before the layer assignment. A split point is a point in a net that connects two or more wire segments. A via location is always considered a split point. The split degree of a split point is the number of wire segments connecting at the split point.

Because a given layout of the CVM problem contains a set of possible vias, a crossing graph, G, for the layout L can be defined and constructed as follows:

$$G = (N, E), \text{ where}$$
$$N = \{v_i| \text{ there is a wire segment } n_i \in L\}$$
$$E = \{(v_i, v_j)| \text{ wire segments } n_i \text{ and } n_j \text{ cross each other}\}$$

Initially, each net is considered as a single wire segment. Depending on the split degree at the via location, a wire segment can be broken into two or more segments by choosing one via along the wire segment. The corresponding crossing graph should be 2-colorable in order to realize the layout on two layers. The algorithm is based on the following theorems [DEO74]:

Theorem 7.7. A graph G is 2-colorable if and ony if G is bipartite.

Theorem 7.8. A graph G is bipartite if and only if there is no odd-length cycle in G.

To verify the 2-colorability of a crossing graph, a depth-first search algorithm can be used in linear time. For a crossing graph G which is not 2-colorable, a set of vias is required to realize segment assignment on two layers. Once a via is chosen on a wire segment, that segment is broken in two. The corresponding vertex in the crossing graph is split into two vertices if the split degree is two at this possible via. This process is continued until the final crossing graph is bipartite. The corresponding layout can then be implemented on two layers. The objective of the algorithm is to choose a set of vias as small as possible, such that the final crossing graph is 2-colorable.

Example: A layout and its corresponding crossing graph are shown in Figure 7.13, where net 1 crosses nets 4 and 5 in the layout (Figure 7.13a). Consequently, node 1 is connected with nodes 4 and 5 in the crossing graph (Figure 7.13b).

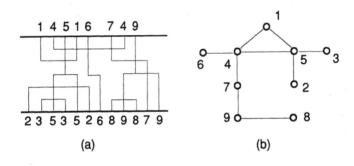

Figure 7.13. Layout and crossing graph

◇

The Layer Assignment Algorithm

According to Theorem 7.8, all the cycles in the crossing graph should be of even length so that the layout can be realized on two layers. Therefore, a set of vias should be selected such that all cycles of odd length in the crossing graph can be broken. Because finding all the odd-length cycles requires exponential time, an efficient heuristic approach for selecting vias is required. In a given crossing graph, the average number of cycles with a length of three is much larger than the number of cycles of odd length greater than three, especially when the crossing graph corresponds to a grid-line-based layout [CHA87]. Because a cycle of odd length greater than three may share some wire segments with a cycle of length three, both cycles may be broken by simply breaking the cycle of length three.

The layer assignment algorithm breaks the cycles of length three first. Then the rest of the cycles of odd length are broken, until all the odd-length cycles are broken. The algorithm proceeds as follows:

1. Generate an initial crossing graph G for a given layout L.

2. Find all cycles of length three in G. Traverse the loops corresponding to the cycles of length three and select a set of vias such that all loops in L are broken. Update crossing graph G.

3. Find a fundamental set of cycles for the current G. Select a via involved in the largest number of fundamental cycles of odd length until the lengths of all fundamental cycles are even. Update G in each iteration. Note that the crossing graph is bipartite after step 3.

4. Use a depth-first-search algorithm to color vertices with two colors in the crossing graph such that no two adjacent vertices have the same colors.

Representing the physical layout relationships among the nets, crossing points, and possible vias, entails constructing a linked list for each net as the data structure. The linked list can be created by traversing the net from the beginning terminal to the ending terminal. All the crossing points and possible vias encountered during traversing are linked to this list. In order to determine which two nets are crossing at a crossing point, a crossing point table should be created to contain this information.

Example: A layout and linked lists are shown in Figure 7.14. Nets are labeled with numbers, and crossing points are labeled with capital letters in the layout L (Figure 7.14a). The possible via locations are labeled with lower case letters, where the wire segments are long enough to place vias between two crossing points. Segments too short to place vias are not labeled. Crossing points A

and B, segment c (possible via location), crossing points C and D, segment d, and crossing point E will be encountered when net 1 is traversed from M_3 to M_4. The corresponding linked list for net 1 is constructed as shown in Figure 7.14b.

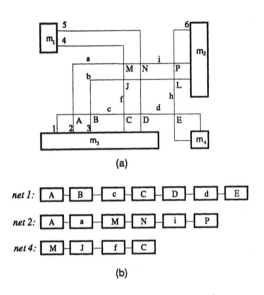

(a)

net 1: [A]—[B]—[c]—[C]—[D]—[d]—[E]

net 2: [A]—[a]—[M]—[N]—[i]—[P]

net 4: [M]—[J]—[f]—[C]

(b)

Figure 7.14. Layout L and part of its linked lists

◇

The time complexity needed to determine all the cycles of length three is $O(|N| \cdot d^2)$, where N is the number of vertices and d is the maximum degree of the vertices in the crossing graph [CHA87]. Because $d = O(N)$, the time complexity is still bounded by $O(|N|^3)$. For each cycle of length three in the crossing graph, the corresponding loop of wire segments in the layout can be determined by traversing the linked lists. Other information, such as possible via locations, can also be retrieved while traversing the lists. Thus, the loop does not have to be traversed. Instead, information can be obtained by traversing the corresponding linked lists for the nets.

For example, there is a corresponding loop in the layout (Figure 7.14a) for the cycle formed by nets 1, 2 and 4 in the crossing graph. The linked lists for nets 1,2 and 4 are traversed. Loop information, including the labeled segments a, f, and c, (possible via locations), can be obtained. After all the loops corresponding to all cycles of length three are traversed, a cover table with a row for each loop and a column for each labeled wire segment can be generated.

A minimum number of labeled wire segments has to be determined such that the vias will be placed to break all the cycles of length three. Because this problem is eqivalent to the subset cover problem, which has been shown to be NP-complete [GAR79], a simple approximation method can be used to select the labeled wire segments in the columns. In this method, the labeled wire segment involved in the most loops is selected. This procedure is repeated until all the cycles are broken. In step 3, the net involved in the largest number of cycles of odd length is identified. The via in this net is located where it will separate the other nets crossing the net into two more-or-less equally sized sets. For example, net 1 is involved in the largest number of odd-length cycles and has two labeled wire segments, c and d. A via on segment c will separate other crossing nets into two sets, {2,3} and {4,5,6}, while a via on segment d will separate other crossing nets into two sets, {2,3,4,5} and {6}. Therefore, wire segment c is a better choice. In step 4, the crossing graph is bipartite. The layer assignment process can then assign segments to layers.

When constraints exist, the algorithm may be modified to ensure that all selected vias satisfy these constraints. For example, the number of vias chosen in some nets may not be greater than a given constant, and the density of vias selected in a given area may be less than a certain degree.

A similar algorithm was presented for the UVM problem [CHA87]. Initially, a crossing graph is constructed. In the second step, all the cycles of length three are generated, and a similar cover table is constructed. The difference is that now each column represents a vertex in the graph. In the third step, an approximation method is used to select a minimum number of columns so that all rows can be covered. The corresponding vertices are deleted from the graph. If the crossing graph is not bipartite, the last step will generate all the fundamental cycles and delete a vertex involved in the maximum number of fundamental cycles of odd length. This step is repeated until the lengths of all the fundamental cycles are even and the reduced graph is bipartite. Experimental results suggest that this algorithm generates a better solution than the maximum independent set approach [HSU83]. For those cases where the number of vertices is not greater than fifteen, the results generated by the algorithm are all optimal.

7.4.2 A 2-CVM algorithm for topological layouts

Naclerio [NAC87B] presented a basic 2-CVM algorithm for topological layouts where two layers are available for routing and the junction degree is limited to three. The algorithm yields a globally optimum solution for the problem. Without loss of generality, each terminal is assumed to be the endpoint of exactly one routing segment in the algorithm. For a given layout L, a graph $G(L) = (N, E)$ is defined by:

$N = C \cup J \cup T$, where C, J, T are the sets of crossings, junctions and terminals in L, respectively

$E = \{(x, y) | x, y \in N$ and there is a subsegment between x and y in $L\}$

Let $loc(x)$ denote the location of the node x and let $seg(x, y)$ be the subsegment in L corresponding to the edge (x, y) in $G(L)$. Let $via(x, y)$ be some legal via located on $seg(x, y)$. A via is assumed to be placed only on a segment, because a via placed on a three-way-split can be shifted slightly to a respective segment, as shown in Figure 7.16 [MOL87]. The derived graph $G(L)$ is planar because no edges intersect.

Example: A layout L and graph $G(L)$ are shown in Figure 7.15a and Figure 7.15b respectively, where junction nodes in (b) correspond to the vias and junctions in (a). Note that the nodes corresponding to the terminals are not shown in the figure; each cycle is labeled with a number.

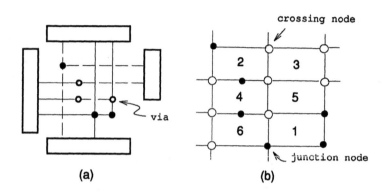

Figure 7.15. Layout L and graph $G(L)$

◇

Consider a cycle consisting of $(n_{i_1}, n_{i_2}), (n_{i_2}, n_{i_3}), \ldots, (n_{i_k}, n_{i_1})$ in $G(L)$. Let n_{i_j} be the node in the cycle and $n_{i_j} \in C$. If the number of node n_{i_j} in the cycle is odd, then the cycle is called an *odd cycle*; otherwise it is called an *even cycle*.

Figure 7.16. Equivalent via locations

Lemma 7.6. Given a topological layout $L = (M, T, W)$, where M, T, W are the sets of modules, terminals, and nets, respectively, a valid topological layout exists without any via if and only if all cycles in $G(L)$ are even.

Proof: Assume that all cycles in $G(L)$ are even. Then all layer changes can be made at the crossing points in L. Thus, a valid topological layout exists without any via.

Conversely, suppose that a valid topological layout L exists without any via. Each cycle consists of a sequence of edges $(n_{i_1}, n_{i_2}), (n_{i_2}, n_{i_3}), \ldots, (n_{i_k}, n_{i_1})$. Consider each pair of edges $(n_{i_{k-1}}, n_{i_k})$ and $(n_{i_k}, n_{i_{k+1}})$. If $n_{i_k} \in C$, $seg(n_{i_{k-1}}, n_{i_k})$ and $seg(n_{i_k}, n_{i_{k+1}})$ cross at $loc(n_{i_k})$. They must be assigned to different layers. If $n_{i_k} \in J$, $seg(n_{i_{k-1}}, n_{i_k})$ and $seg(n_{i_k}, n_{i_{k+1}})$ form a junction at $loc(n_{i_k})$. Because $loc(n_{i_k})$ cannot be a via (a via is not located on a junction point), $seg(n_{i_{k-1}}, n_{i_k})$ and $seg(n_{i_k}, n_{i_{k+1}})$ must be assigned to the same layer. Thus, the layers can only be changed at crossings, and the total number of such changes must be even. Otherwise, the result is an invalid layout. □

If a via is placed on a subsegment $seg(x, y)$, a layer change can be made on this subsegment. The routing segment which includes $seg(x, y)$ is partitioned into two routing segments at the via, as shown in Figure 7.17.

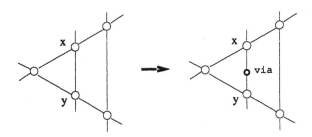

Figure 7.17. Inserting a via on $seg(x, y)$

The corresponding edge in $G(L)$ can then be removed and two cycles sharing that edge can be merged. If the remaining graph is odd-cycle free, then no further vias are required by Lemma 7.6. Note that the number of odd cycles is always even and an odd cycle can only be eliminated by merging it with another odd cycle. If an odd cycle does not have an adjacent odd cycle, it must merge with several even cycles before merging with a particular odd cycle. In Figure 7.15b, odd cycle 2 has to merge with even cycle 3 before merging with odd cycle 5.

Definition 7.11. A set of edges is called an *odd cycle cover* (OCC) if removal of the edges leaves $G(L)$ free of odd cycles.

Definition 7.12. An OCC of minimum cardinality is called a *minimum odd cycle cover* (MOCC).

Lemma 7.7. Given a valid topological layout L, if F is a MOCC for $G(L)$, then a valid topological layout exists such that the set of vias, $V = \{via(x,y)|(x,y) \in F\}$, is of minimum cardinality.

Proof: Because there is a one-to-one correspondence between the edges in $G(L)$ and the via candidates in L, the lemma is valid according to the definition of a MOCC. □

The problem of finding a MOCC from the planar graph $G(L)$ can be solved by determining pairs of odd cycles such that the total number of edges that must be cut in each path between two cycles to be merged is minimum [HAD75]. These edges are common edges between each two adjacent cycles while traversing along the path from one odd cycle to another. For example, a path is traversed in Figure 7.15b, beginning at cycle 2, then entering cycle 3, and finaly ending at cycle 5.

A geometric dual graph $G_d = (N_d, E_d)$ is generated from $G(L)$, where each node in N_d corresponds to a face (cycle) and each edge in E_d corresponds to two adjacent faces in $G(L)$. Let map be a mapping between edge $(x,y) \in G_d$ and edge $(r,s) \in G(L)$ be cut by (x,y). The node in G_d corresponding to an odd cycle in $G(L)$ is called an *odd node*. The number of odd nodes is even because the number of odd cycles is even. Thus, removal of an edge in $G(L)$ corresponds to contracting an edge in G_d. When two nodes are merged to contract an edge in G_d, the new node is even only if these two nodes are odd.

The problem of finding a MOCC can now be solved by finding a set of edges with minimum cardinality in G_d whose contracting frees G_d of odd nodes. This set of edges is called an *odd node pairing*, and the set with minimum cardinality is called a *minimum odd node pairing*. Hadlock [HAD75] showed that a set of edges is an OCC of a planar graph G if and only if the corresponding edge set is an *odd node pairing* for the dual graph G_d of G. He also showed that a *minimum odd node pairing* consists of a set of paths between pairs of odd nodes such that edges are disjoint in the paths, and each node is the endpoint of exactly one such path.

Thus, a *minimum odd node pairing* can be found by solving a maximum weight matching problem. First, a complete graph $G_c = (N_{odd}, E_c)$ is constructed from G as follows:

$$N_{odd} = \{n_d | n_d \text{ is an odd node in } G_d\}$$
$$E_c = \{(n_i, n_j) | n_i, n_j \in N_{odd}, i \neq j\}$$

For each pair of nodes x and y in N_{odd}, let P_{x-y} be the path between x and y in graph G_d, which consists of all edges along the shortest path between x and y in G_d. A set of via candidates V_{x-y} corresponding to the path P_{x-y} is defined as:

$$V_{x-y} = \{via(map(a, b)) | (a, b) \in P_{x-y}\}$$

For each edge $(a, b) \in P_{x-y}$, the corresponding cutting edge (r, s) in $G(L)$ is obtained first by $map(a, b)$, and then the via location is found by $via(r, s)$. For each edge $(x, y) \in G_c$, a weight is defined as $W(x, y) = w - |V_{x-y}|$, where w is some large constant.

A maximum weight matching problem for G_c is finding a subset $E_x \subseteq E_c$ such that no node can be the endpoint of more than one edge in E_x, and $\sum_{(x,y) \in E_x} W(x, y)$ is maximum. The maximum weight matching problem can be solved in polynomial time, using the algorithms developed in [EDM60, GAB74, LAW76]. Once the E_x has been found, an MONP can be determined by $\cup_{(x,y) \in E_x} P_{x-y}$, and a minimum cardinality set of vias is given by $\cup_{(x,y) \in E_x} V_{x-y}$. The basic algorithm proceeds as follows:

- Generate the graph $G(L) = (N, E)$. Label each edge (x, y) with a pointer $via(x, y)$ to a via location on the segment $seg(x, y)$.

- Generate the dual graph $G_d(N_d, E_d)$ from $G(L)$. Find the set of odd nodes, $N_{odd} \subseteq N_d$.

- Generate the complete graph $G_c = (N_{odd}, E_c)$. Assign a weight $W(x, y)$ for each edge $(x, y) \in E_c$ and $W(x, y) = w - |V_{x-y}|$, where w is some large constant and $V_{x-y} = \{via(map(a, b)) | (a, b) \in P_{x-y}\}$.

- Find a maximum weight matching $E_x \subseteq E_c$.

- Determine a minimum cardinality set of vias, $V = \cup_{(x,y) \in E_x} V_{x-y}$.

- Make a layer assignment.

Example: Figure 7.18a shows a complete graph G_c with edge weights constructed from the dual graph G_d. The corresponding graph $G(L)$ is shown in
Figure 7.15b. Figure 7.18b shows a path $P_{x_2-x_5} = \{(x_2,x_3),(x_3,x_5)\}$, where
x_2, x_3 and x_5 are nodes in G_d. The corresponding set of via candidates is $V_{x_2-x_5} = \{via(r_2,r_3), via(r_3,r_5)\}$.

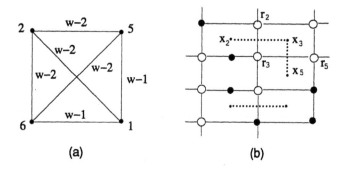

(a)   (b)

Figure 7.18. Complete graph G_c and path $P_{x_2-x_5}$

◇

Time Complexity

All of the steps in the algorithm can be performed in polynomial time. The first
two steps can be conducted in $O(n^2)$, where n is the number of the routing segments. In the last step, each subsegment is traversed exactly once, so the layer
assignment can be accomplished in $O(n^2)$. Determining a minimum cardinality set of vias can be carried out in linear time with respect to the number of
vias, which cannot exceed the number of junctions in a given layout. Thus, the
time is bounded by n. The computation of edge weights can be conducted in

$O(|E_d| \cdot |N_{odd}|)$ by using a breadth-first-search [AHO74, EVE79], where E_d and N_{odd} are the sets of edges and odd nodes in G_d, respectively. Because $|E_d| = O(n^2)$, the time for the computation of edge weights is bounded by $O(|N_{odd}| \cdot n^2)$. Finding a maximum weight matching on the graph G_c can be conducted in $O(|N_{odd}|^3)$ [GAB74, LAW76]. Because the number of odd nodes is bounded by $O(n)$ [CHE83], the total time complexity for the algorithm is $O(n^3)$, where n is the number of routing segments.

Naclerio [NAC87B] has implemented the algorithm in C programming language. He applied the algorithm to a number of examples of real channel routing problems that were first routed using the channel routing algorithm described in [RIM87]. In all cases, the time required to obtain optimum layer assignments is less than the upper bound of $O(n^3)$. The number of vias required for these examples is reduced by between 14 percent and 41 percent of the number required by the original layer assignment, where all vertical routing segments were placed on one layer, and all horizontal routing segments were placed on the other.

7.5   References

[AHO74]   Aho, A. V., Hopcroft, J. E. and Ullman, J. D., *The Design and Analysis of Computer Algorithms*, Addison-Wesley Publishing Company, 1974.

[CHA87]   Chang,K. C. and Du,D. H., "Efficient Algorithms for Layer Assignment Problem," *IEEE Trans. on Computer-Aided Design*, vol. CAD-6, no. 1, pp. 67-78, Jan. 1987.

[CHE82]   Chen, R. W., Kajitani, Y. and Chan, S.P., "Topological Considerations of the Via Minimization Problem for Two-Layer PC Board," in *Proc. ISCAS*, pp. 968-971, 1982.

[CHE83]   Chen, R-W., Kajitani, Y. and Chan, S-P., "A Graph-Theoretic Via Minimization Algorithm for Two-Layer Printed Circuit Boards," *IEEE Trans. Circuits and Systems*, vol. CAS-30, no. 5, pp. 284-299, May 1983.

[CHI85]   Chiba, N., Nishizeki, T., Abe, S. and Ozawa, T., " A Linear Time Al-
            gorithm for Embedding Planar Graphs Using PQ-Trees," *J. Comput.
            System Sci.*, 30,1985, pp. 54-76.

[CHO89]   Choi, H., Nakajima, K. and Rim, C. S., "Graph Bipartization and
            Via Minimization," *SIAM J. Disc. Math.*, vol. 2, no. 1, pp. 38-47,
            Feb. 1989.

[CON91]   Cong, J. and Liu, C. L., "On the k-Layer Planar Subset and Topolog-
            ical Via Minimization Problems," *IEEE Trans. on Computer-Aided
            Design*, vol. 10, no. 8, pp. 972-981, Aug. 1991.

[DEO74]   Deo, N., *Graph Theory with Application to Engineering and Computer
            Science.* Englewood Cliffs, NJ: Prentice-Hall, 1974.

[DEO89]   Deogun, J. S. and Bhattacharya, B. B., "Via Minimization in VLSI
            Routing with Movable Terminals," *IEEE Trans. on Computer-Aided
            Design*, vol. 8, no. 8, pp. 917-920, Aug. 1989.

[DU84]   Du, H. C. and Chang, K. C., "A New Approach for Layer Assignment
            Problem," Tech. Rep., TR-84-20, Computer Science Dept., Univ.
            Minnesota, Minneapolis, MN, 1984.

[EDM60]   Edmonds, J., "Maximum Matching and a Polyhedron with 0-1 Ver-
            tices," *J. Res. Nat. Bur. Standards*, vol. 69B, no. 1-2, pp. 125-130,
            June 1960.

[EVE71]   Even, S. and Itai, A., "Queues, Stacks, and Graphs," in *Theory of
            Machines and Computations*, Z. Kohavi and A. Paz, eds. New York:
            Academic, 1971, pp. 71-86.

[EVE79]   Even, S., *Graph Algorithms.* Rockville, MD: Computer Science Press,
            1979.

[FAR48]   Fary, I., "On Straight Line Representation of Planar Graphs," *Acta
            Sci. Math. (Szeged)*, vol. 11, pp. 229-233, 1948.

[GAB74]   Gabow, H. N., "Implementation of Algorithms for Maximum Match-
            ing on Nonbipartite Graphs," Department of Computer Science, Stan-
            ford University, Stanford, CA, Ph.D. Dissertation, 1974.

[GAR76]   Garey, M. R. and Johnson, D. S., "Some Simplified NP-complete
            Graph Problems," *Theor. Comput. Sci.*, 1 (1976), pp. 237-267.

[GAR79]   Garey, M. R. and Johnson, D. S., *Computers and Interactability: A Guide to the Theory of NP Completeness*, W. H. Freeman, San Francisco, 1979.

[HAD75]   Hadlock, F., "Finding a Maximum Cut of a Planar Graph in Polynomial Time," *SIAM J. Comput.*, vol. 4, no. 5, pp. 221-225, Sept. 1975.

[HOP79]   Hopcroft, J. E. and Ullman, J. D., *Introduction to Automata Theory, Languages, and Cmputation*, Addison-Wesley Publishing Company, 1979.

[HSU83]   Hsu, C. P., "Minimum-Via Topological Routing," *IEEE Trans. Computer-Aided Design*, vol. CAD-2, no. 4, pp. 235-246, Oct. 1983.

[KRI79]   Krishnamoorthy, M. S. and Deo, N., "Node-Deletion NP-complete Problems," *SIAM J. Comput.*, 8 (1979), pp. 619-625.

[LAW76]   Lawler, E. L., *Combinatorial Optimization: Networks and Matroids.* New York: Holt, Rinehart, and Winston, 1976.

[LIC82]   LiChtenstein, D., "Planar Formulae and Their Uses ," *SIAM J. Comput.*, Vol. 11, no. 2, pp. 329-343, May 1982.

[MAR84]   Marek-Sadowska, M., "An Unconstrained Topological Via Minimization Problem for Two-Layer Routing," *IEEE Trans. on Computer-Aided Design*, CAD-3, pp. 184-190, July 1984.

[MOL87]   Molitor,P., "On the Contact-Minimization-Problem, " in *Proc. 4th Annual Symp. on Theoretical Aspects of Computer Science*, Passau, West Germany, Feb. 1987; *Lecture Notes in Computer Science* 247, Springer-Verlag, New York, Berlin, 1987, pp.420-431.

[NAC87A] Naclerio, N. J., Masuda, S. and Nakajima, K., "Some NP-Complete Results for Via Minimization," in *Proc. 1987 Conf. on Information Sciences and Systems*, Baltimore, MD, Mar. 1987, pp. 611-616.

[NAC87B] Naclerio, N. J., "Via Minimization for IC and PCB Layouts," Department of Electrical Engineering, University of Maryland, College Park, MD, Ph.D. Dissertation, 1987.

[PIN82]   Pinter, R. Y., "Optimal Layer Assignment for Interconnect," in *Proc. ISCAS*, 1982, pp. 398-401.

[RIM87]   Rim, C. S. and Nakajima, K.,"An Efficient Channel Routing Algo-
            rithm for Two and Three Layers," unpublished report, System Re-
            search Center, University of Maryland, College Park, MD, 1987.

[SAR89]   Sarrafzadeh,M. and Lee,D. T., "A New Approach to Topological Via
            Minimization," *IEEE Trans. on Computer-Aided Design*, vol. 8, no.
            8, pp. 890-900, Aug. 1989.

[STA90]   Stallmann, M., Hughes, T. and Liu,W., "Unconstrained Via Min-
            imization for Topological Multilayer Routing," *IEEE Trans. on
            Computer-Aided Design*, vol. 9, no. 9, pp. 970-980, Sep. 1990.

[TAM87]   Tammassia,R., "On Embedding a Graph in the Grid with the Min-
            imum Number of Bends," *SIAM J. Comput.*, vol. 16, no. 3, pp.
            421-444, June 1987.

Chapter 8

A Solution for Steiner's Problem

Yeun Tsun Wong and Michael Pecht

Associated with the rapid improvement of fine-line etching techniques, both the wire density and the number of routing layers have been skyrocketing. Providing new routing techniques for minimizing the number of layers (or the routing area in each layer) and the number of vias is getting more and more important. Because the length of minimum rectilinear Steiner trees (MRSTs) is less than that of minimum spanning trees by about 20% on average, and because there are many choices for connecting a set of signal terminals with MRSTs (see Section 2.2.2 of Chapter 2), Steiner's problem with rectilinear distance dominates almost all important routing problems: routing a net with a shortest length, minimizing the number of layers (or routing area), and minimizing the number of vias.

Steiner's problem with rectilinear distance is an infamous NP problem. It was usually reduced to subproblems (restricting the number of nodes by less than 6, or restricting node locations on the perimeter of a rectangle), or was approximated by improving a minimum spanning tree to reduce tree length by about 8% on average. However, the approximations do not provide alternative trees. Therefore, it is critical to provide Steiner's problem a general solution, which can be computed in a polynomial time, except for the worse and few nearly worse cases.

In order to find the general solution of Steiner's problem with rectilinear distance, this chapter discusses a rejection rule, that rejects trees which cannot be MRSTs until only MRSTs are left. This rejection rule is based on a connected

graph called an edge-set tree, which is generated by expanding each edge of a
rectilinear Steiner tree to an edge-set (a set of all edges connecting the same pair
of vertices) enclosed exactly by a rectangle whose half-perimeter is equal to the
length of an edge. This rejection rule states that a redundancy exists between
two adjacent edge-sets, if more than one point exists in the intersection of two
rectangles exactly enclosing the adjacent edge-sets. Any edge-set trees that can
be converted to an edge-set tree in which a redundancy exists are also rejected.
Thus, Steiner's problem with rectilinear distance can be transformed to a problem
finding all edge-set trees containing MRSTs.

In order to ease the reading, complicative proofs are omitted, and only those
theorems and corollaries needed to find no-redundancy edge-set trees in a polynomial time are presented. This chapter first gives the definitions pertaining to
Steiner points, edge-sets and edge-set trees, and discusses properties of Steiner
points in an no-redundancy edge-set tree. Based on the previous discussion, next
studies the structure of a clique, which consists of dynamic Steiner points and
their adjacent edge-sets. Then develops graph arithmetics for generating a no-
redundancy edge-set tree, and generating other no-redundancy edge-set trees, by
replacing edge-sets and cliques of the no-redundancy edge-set tree with other
edge-sets and cliques. Finally describes the general solution contained by all no-
redundancy edge-set trees, and provides procedures for finding no-redundancy
edge-set trees, as well as an example of finding the general solution for 10 nodes.

8.1   Basic Definitions

Let (x_r, y_r) be the point r, with x_r and y_r as horizontal and vertical coordinates,
and let P_{pq} be a set of points enclosed by rectangle R_{pq}, with four vertices p, q,
(x_p, y_q) and (x_q, y_p). P_{pq} can be expressed as

$$P_{pq} = \{r = (x_r, y_r) \mid \ x_r = min(x_p, x_q), ..., max(x_p, x_q),$$
$$y_r = min(y_p, y_q), ..., max(y_p, y_q)\} \qquad (8.1)$$

where min and max are functions for selecting a minimum element and a maximum element from a set, respectively. The perimeter, $l_{R_{pq}}$, of R_{pq} can be measured by the number of points lying on the perimeter, and the ith rectilinear
edge from p to q, $e_{pq}(i)$, can be expressed as a set of $l_{R_{pq}}/2 + 1$ adjacent points

$\{p, r_1^{(i)}, r_2^{(i)}, ..., r_{(l_{R_{pq}}-1)/2}^{(i)}, q\} \subset P_{pq}$. Therefore, the length of $e_{pq}(i)$, $l_{e_{pq}(i)}$, equals $l_{R_{pq}}/2 = |x_p - x_q| + |y_p - y_q|$. If an arbitrary point in a set of points, denoted as M_p, can be selected as p, then M_p is called the domain of p, and $p \in M_p$ also means p can be moved in the area occupied by M_p. Based on the domain of a point, a set of all edges from p to q, E_{pq}, can be expressed as:

$$E_{pq} = \{e_{pq}(i)| \; p \in M_p, \; q \in M_q, \; l_{e_{pq}(i)} = \frac{1}{2} l_{R_{pq}}, \; i = 1, ..., |E_{pq}|\} \qquad (8.2)$$

where $|E_{pq}| = C(|x_p - x_q| + |y_p - y_q|, |x_p - x_q|)$ is the total number of edges from p to q. Note that $\cup_{i=1}^{|E_{pq}|} e_{pq}(i) = P_{pq}$. P_{pq} is called a point-set of E_{pq}. When $x_p = x_q$ or $y_p = y_q$, E_{pq} is called a straight edge-set that contains a set of straight edges. E_{pq} is a straight edge-set, if and only if every edge in E_{pq} is a straight edge. When $x_p \neq x_q$ and $y_p \neq y_q$, E_{pq} is called a bend edge-set. E_{pq} is a bend edge-set, if and only if one edge in E_{pq} is a bend edge. A straight edge may be contained in bend edge-set E_{pq} if $|M_p| > 1$ or $|M_q| > 1$, and all edges in bend edge-set E_{pq} are bend edges if $|M_p| = |M_q| = 1$.

Example: Figure 8.1a shows rectangle R_{pq} with p and q as vertices. Figure 8.1b shows the point-set of E_{pq}, P_{pq}, which is a set of all points enclosed by R_{pq}. Figure 8.1c shows E_{pq}, which is a set of all edges connecting p and q. Each of the six edges shown in Figure 8.1c connects five points. Because there are a total of eight points on the perimeter of R_{pq}, the length of every edge from p to q is four. Because a bend edge exists in E_{pq}, E_{pq} is a bend edge-set. If $M_p = \{(0,0), (1,0), (2,0)\}$ and $p = (2,0)$, then a vertical straight edge with $(2,0)$ and $(2,2)$ as vertices is contained in E_{pq}. However, E_{pq} is still a bend edge-set.

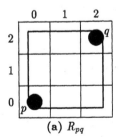 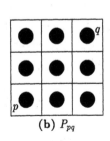 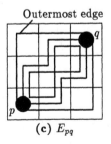

(a) R_{pq} (b) P_{pq} (c) E_{pq}

Figure 8.1. A rectangle, a point-set (a set of points in the rectangle), and an edge-set (a set of edges enclosed by the rectangle)

Because R_{pq} is exactly occupied by all edges from p to q, and because every edge contained by E_{pq} consists of $(l_{e_{pq}(i)}+1)$ adjacent points from p to q in R_{pq}, finding R_{pq} is equivalent to finding E_{pq} without regard to a specific edge. Therefore, $e_{pq}(i)$ is abbreviated as e_{pq}, and the length of E_{pq} is denoted as $l_{E_{pq}}$, which equals $l_{e_{pq}}$.

When $p = q$, P_{pq} and E_{pq} are reduced to a point. When $|M_p| = 1$, p is either a given node or a fixed Steiner point. When $|M_p| > 1$, p is a dynamic Steiner point. When M_p is a set of points lying on a straight edge, then p is moved along either the horizontal or the vertical direction, and p is a one-dimensional Steiner point. When p can be moved along both horizontal and vertical directions, p is a two-dimensional Steiner point.

Based on the domain of each end-vertex, edge-sets can be classified into fixed types and a moving types. When $|M_p| = |M_q| = 1$, E_{pq} is a fixed edge-set. The number of all edges contained in E_{pq}, $A_0(p,q)$, is

$$A_0(p,q) = C(|x_p - x_q| + |y_p - y_q|, |x_p - x_q|) \tag{8.3}$$

Example: In Figure 8.1c, $p = (0,0)$ and $q = (2,2)$. Therefore, $A_0(p,q) = C(2+2,2) = 6$ and $l_{E_{pq}} = l_{e_{pq}} = |x_p - x_q| + |y_p - y_q| = |0-2| + |0-2| = 4$.
⋄

When $|M_p| \geq 1$ and $|M_q| = 1$, E_{pq} is called a one-vertex moving edge-set, and it represents $|M_p|$ different fixed edge-sets. When $|M_p| \geq 1$, $|M_q| \geq 1$, E_{pq} is called a two-vertex moving edge-set, in which the number of different fixed edge-sets is $|M_p||M_q| - C(|M_p \cap M_q|, 2) = |M_p||M_q| - |M_p \cap M_q|(|M_p \cap M_q| - 1)/2$, if p and q can be moved independently. In all fixed edge-sets included in moving edge-set E_{pq}, the one with the minimum length is called the shortest edge-set, denoted as E_{pq}^{min}, and the one with the maximum length is called the longest edge-set, denoted as E_{pq}^{max}. The average length of E_{pq} can be obtained by setting $p = (\bar{x}_p, \bar{y}_p)$ and $q = (\bar{x}_q, \bar{y}_q)$: $\bar{l}_{E_{pq}} = |\bar{x}_p - \bar{x}_q| + |\bar{y}_p - \bar{y}_q|$, where $\bar{x}_p = \sum_{i=1}^{|M_p|} x_{p_i}$, $\bar{y}_p = \sum_{i=1}^{|M_p|} y_{p_i}$, $\bar{x}_q = \sum_{i=1}^{|M_q|} x_{q_i}$, $\bar{y}_q = \sum_{i=1}^{|M_q|} y_{q_i}$, and (x_{r_i}, y_{r_i}) ($r = p$ or q) is a point in the ith location of M_r.

Example: If $M_p = \{(0,0), (0,1), (1,0), (1,1)\}$ and $M_q = \{(1,1), (2,1), (1,2), (2,2)\}$, then $M_p \cap M_q = (1,1)$. The number of fixed edge-sets included by E_{pq} is $|M_p||M_q| - |M_p \cap M_q|(|M_p \cap M_q| - 1)/2 = 4 \times 4 - 1(1-1)/2 = 16$. Because $\bar{x}_p = (0+0+1+1)/4 = 1/2$, $\bar{y}_p = (0+1+0+1)/4 = 1/2$, $\bar{x}_q = (1+2+1+2)/4 = 3/2$ and

$\bar{y}_q = (1 + 1 + 2 + 2)/4 = 3/2$, $\bar{l}_{E_{pq}} = 2$. When $p = q = (1, 1)$, E_{pq} becomes E_{pq}^{min}. When $p = (0, 0)$ and $q = (2, 2)$, E_{pq} becomes E_{pq}^{max}.

◇

When $x_p = x_q$ and M_p and M_q are two sets of points lying on two opposite sides of R_{ab}, where $a = (x_a, y_p)$ and $b = (x_b, y_q)$, straight edge $e_{pq} \in E_{pq}$ can be horizontally moved in R_{ab} without changing $l_{e_{pq}}$ (see Figure 8.2a). Under this circumstance, p and q are one-dimensional Steiner points, and E_{pq} is called a horizontal parallel moving edge-set. Note that the locations of p and q are dependent. If $x_a = x_b$, then M_p and M_q are two points, and E_{pq} is reduced to a fixed straight edge. Similarly, a vertical parallel moving edge-set exists (see Figure 8.2b). The fixed edge-set, the one-vertex moving edge-set and the parallel moving edge-set — special cases of the two-vertex moving edge-set — simplify the expression of the interconnections among given nodes.

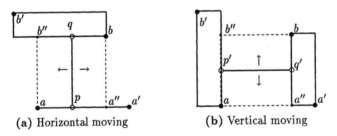

(a) Horizontal moving   **(b)** Vertical moving

Figure 8.2. Parallel moving edge-sets

Example: Figure 8.2a shows that edge-sets $E_{aa'}$ and $E_{bb'}$ are connected by edge-set E_{pq}, where p can be moved from a to a'', q can be moved from b'' to b, and $x_p = x_q$. Therefore, $M_p = P_{aa''} = e_{aa''}$, $M_q = P_{b''b} = e_{b''b}$, p and q are dependent, e_{pq} is a vertical edge, and E_{pq} is a horizontal parallel moving edge-set.

◇

If E_{pq} is a parallel moving edge-set connecting edge-sets $E_{aa'}$ and $E_{bb'}$, where $P_{aa'} \cap P_{bb'} = \emptyset$, $P_{pq} \cap P_{aa'} = p$, and $P_{pq} \cap P_{bb'} = q$, then M_p and M_q are called overlaps of edge-sets $E_{aa'}$ and $E_{bb'}$, or of point-sets $P_{aa'}$ and $P_{bb'}$ on the direction perpendicular to E_{pq}. Defining $\natural$ as an overlap operator, the overlap operation of $E_{aa'}$ and $E_{bb'}$, or of $P_{aa'}$ and $P_{bb'}$, is defined as:

$$E_{aa'} \natural E_{bb'} = P_{aa'} \natural P_{bb'} = \begin{cases} \{M_p, M_q| \; x_p = x_q \; or \; y_p = y_q, \; |M_p| = |M_q| \geq 1\} \\ \emptyset \quad (\text{when no overlap}) \end{cases} \quad (8.4)$$

If $|M_p| = |M_q| = 1$, E_{pq} is a fixed straight edge-set.

Example: In Figure 8.2a, e_{pq} is a horizontal parallel moving edge, and $E_{aa'}$ and $E_{bb'}$ are horizontal overlap edge-sets with the overlaps $M_p = e_{aa''}$ and $M_q = e_{bb''}$. The overlap operation is expressed as $E_{aa'} \natural E_{bb'} = P_{aa'} \natural P_{bb'} = e_{aa'} \natural e_{bb'} = \{M_p, M_q | \ x_p = x_q, |M_p| = |M_q| > 1\}$. In Figure 8.2b, $E_{ab'}$ and $E_{a'b}$ are vertical overlap edge-sets, and $E_{ab'} \natural E_{a'b} = \{M_{p'}, M_{q'} | \ y_{p'} = y_{q'}, |M_{p'}| = |M_{q'}| > 1\}$.

◇

Theorem 8.1. The minimum total length needed to connect a, a', b and b' is $l_{E_{aa'}} + l_{E_{bb'}} + l_{E_{pq}}$, if $E_{aa'} \natural E_{bb'} = \{M_p, M_q | \ x_p = x_q \text{ or } y_p = y_q, 1 \leq |M_p| = |M_q| \leq l_{E_{pq}}\}$.

Proof: In Figure 8.2a, $|M_p| = |M_q| = l_{E_{aa''}} = l_{E_{bb''}}$, because $M_p = P_{aa''} = e_{aa''}$ and $M_q = P_{bb''} = e_{bb''}$ are two opposite sides of R_{ab}. Because $l_{E_{a'a''}}$ and $l_{E_{b'b''}}$ are independent of the movements of p and q, the minimum length for connecting a, a', b and b' is determined by the total length of three of four sides of R_{ab}. Therefore, $l_{E_{aa'}} + l_{E_{bb'}} + l_{E_{pq}} = l_{E_{aa''}} + l_{E_{a'a''}} + l_{E_{bb''}} + l_{E_{b'b''}} + l_{E_{pq}}$ is the minimum total length of edge-sets connecting a, a', b and b', if $1 \leq |M_p| \leq l_{E_{pq}}$. □

Example: Figures 8.2a and 8.2b demonstrate that a, a', b and b' can be connected with the minimum length by two ways, because $l_{E_{pq}} = |M_p|$ or $l_{E_{p'q'}} = |M_{p'}|$. Figure 8.3 presents an example of Theorem 8.1. Because $l_{E_{pq}} > |M_p| = l_{e_{aa''}}$ in Figure 8.3a, the total length of $E_{aa'}$, $E_{bb'}$ and E_{pq} is the minimum length needed to connect a, a', b and b'. In Figure 8.3b, $l_{E_{ab'}} + l_{E_{a'b}} + l_{E_{p'q'}}$ is longer than the minimum length needed to connect a, a', b and b', because $l_{E_{p'q'}} < |M_{p'}|$.

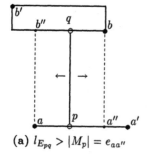 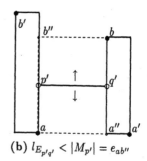

(a) $l_{E_{pq}} > |M_p| = e_{aa''}$ (b) $l_{E_{p'q'}} < |M_{p'}| = e_{ab''}$

Figure 8.3. The total length needed to connect four vertices of two overlap edge-sets

◇

Corollary 8.1.1. The minimum length needed to connect a, a', b and b' is either $l_{E_{aa'}} + l_{E_{bb'}} + l_{E_{pq}}$ or $l_{E_{ab'}} + l_{E_{a'b}} + l_{E_{p'q'}}$, if and only if $E_{aa'} \natural E_{bb'} = \{M_p, M_q \mid x_p = x_q$ or $y_p = y_q, |M_p| = |M_q| = l_{E_{pq}}\}$ or $E_{ab'} \natural E_{a'b} = \{M_{p'}, M_{q'} \mid x_{p'} = x_{q'}$ or $y_{p'} = y_{q'}, |M_{p'}| = |M_{q'}| = l_{E_{p'q'}}\}$.

If $P_{aa'} \natural P_{bb'} = \emptyset$, where a' and $p \in P_{bb'}$ are the nearest points between $P_{aa'}$ and $P_{bb'}$, $P_{aa'}$ can be horizontally or vertically extended to P_{ap} such that $E_{ap} \natural E_{bb'} = \{M_p, M_q \mid x_p = x_q$ or $y_p = y_q, |M_p| = |M_q| = 1\}$. According to Theorem 8.1, the minimum length needed to connect a, p, b and b' is $l_{E_{ap}} + l_{E_{pq}} + l_{E_{bb'}}$. Because $l_{E_{a'q}} = l_{e_{a'p}} + l_{E_{pq}}$, $E_{a'q}$ is the edge-set connecting $E_{aa'}$ and $E_{bb'}$ with the minimum length.

Corollary 8.1.2. The only edge-set connecting $E_{aa'}$ and $E_{bb'}$ with the minimum length is a bend edge-set that connects two nearest points between $P_{aa'}$ and $P_{bb'}$, if and only if $E_{aa'} \natural E_{bb'} = \emptyset$.

If the bend edge set described in Corollary 8.1.2 is degenerated to a point, then Corollary 8.1.3 can be obtained. If two given points overlay, then Corollary 8.1.4 can be obtained.

Corollary 8.1.3. The minimum total length needed to connect a, a', b and b' is $l_{E_{aa'}} + l_{E_{bb'}}$, if $|P_{aa'} \cap P_{bb'}| = 1$.

Corollary 8.1.4. The minimum total length needed to connect a, a' and b is $l_{E_{aa'}} + l_{E_{a'b}}$, if and only if $a' \in P_{ab}$.

According to Corollary 8.1.4, if $a' \in P_{ab}$ is a Steiner point, the number of edges contained in E_{ab} will be reduced. Thus, the definition of the edge-set is violated. Furthermore, if $a' \notin P_{ab}$, then $l_{E_{aa'}} + l_{E_{a'b}}$ is not the minimum total length needed to connect a, a' and b. A Steiner point connected to only two points is called a fake Steiner point and must be removed.

Corollary 8.1.5. If $a' = P_{aa'} \cap P_{a'b}$ is a fake Steiner point, then a' can be removed by replacing $\{E_{aa'}, E_{a'b}\}$ with E_{ab}.

Example: If fake Steiner point f is placed at location $(1,1)$ in the rectangle R_{pq} in Figure 8.1a, the two outermost edges contained in E_{pq} in Figure 8.1c disappear. If f is placed at the location $(0,2)$, only the upper outermost edge exists. If f is outside R_{pq}, the length needed to connect p and q through f is longer than the length of an edge from p to q.

◇

Replacing edges with edge-sets in the graph expression of a tree, an edge-set tree with n_T edge-sets can be expressed as a connected graph:

$$
\begin{aligned}
T &= (E_T, V_T) \\
&= (E_T, V_n \bigcup V_s(T)) \\
&= \{t_j = (E(t_j), V_T) | \ j = 1, ..., N_t\}
\end{aligned}
\tag{8.5}
$$

where $E_T = \{E_{p_i q_i} | \ i = 1, ..., n_T\}$ is a set of all edge-sets in T, $V_T = \cup_{i=1}^{n_T} \{p_i, q_i\}$, V_n is a set of given nodes, $V_s(T)$ is a set of Steiner points in T, $E(t_j) = \{e_{p_i q_i} \in E_{p_i q_i} | \ i = 1, ..., n_T\}$ is a set of edges of tree t_j, and N_t is the total number of different trees contained in T. Denoting l_T as the length of T, and l_t as the length of t, $l_{t_j} = \sum_{i=1}^{n_T} l_{e_{p_i q_i}}$ and $\bar{l}_T = \sum_{i=1}^{n_T} \bar{l}_{e_{p_i q_i}}$. Because T is a connected graph, $P_{p_i q_i} \cap (\cup_{j=1, \neq i}^{n_T} P_{p_j q_j}) \neq \emptyset$. As is the case with a point-set of an edge-set, the set of points occupied by T is called a point-set of T, and is denoted as $P(T)$:

$$
P(T) = \bigcup_{i=1}^{n_T} P_{p_i q_i}
\tag{8.6}
$$

If $T = E_{pq}$, then $P(T) = P(E_{pq}) = P_{pq}$.

Example: Figures 8.4a and 8.4b show edge-set trees $T_a = (E_{T_a}, V_{T_a})$ and $T_b = (E_{T_b}, V_{T_b})$, respectively, where $E_{T_a} = \{E_{n_1 n_2}, E_{n_2 s_1}, E_{s_1 n_3}, E_{s_1 s_2}, E_{s_2 n_4}, E_{s_2 n_5}\}$, $V_{T_a} = V_n \cup V_s(T_a) = \{n_i | \ i = 1, ..., 5\} \cup \{s_1, s_2\} = \{n_i, s_j | \ i = 1, ..., 5; \ j = 1, 2\}$, $E_{T_b} = \{E_{n_i n_{i+1}} | \ i = 1, ..., 4\}$, and $V_{T_b} = V_n = \{n_i | \ i = 1, ..., 5\}$. Every edge-set in T_a or T_b is directly connected to at least one other edge-set in the same edge-set tree. In T_a, edge-sets $E_{n_2 s_1}$, $E_{n_3 s_1}$, $E_{n_4 s_2}$, and $E_{n_5 s_2}$ are one-vertex moving edge-sets, and $E_{s_1 s_2}$ is a parallel moving edge-set, where $x_{s_1} = x_{s_2} = 2, 3, 4$. All moving edge-sets in T_a are dependent on the locations of s_1 and s_2. When $x_{s_1} = x_{s_2} = 3$, $l_{E_{n_2 s_1}} = 3$, $l_{E_{n_3 s_1}} = l_{E_{n_5 s_2}} = 2$, and $l_{E_{n_4 s_2}} = 1$. When $e_{s_1 s_2}$ is moved to the left $(x_{s_1} = x_{s_2} = 2)$, $l_{E_{n_2 s_1}} = l_{E_{n_4 s_2}} = 2$, $l_{E_{n_3 s_1}} = 3$, and $l_{E_{n_5 s_2}} = 1$. Although $E_{s_1 s_2}$ is moved, the length of T_a always equals $l_{E_{n_1 n_2}} + \sum_{i=2}^{3} l_{E_{n_i s_1}} + l_{E_{s_1 s_2}} + \sum_{i=4}^{5} l_{E_{n_i s_2}} = 12$.

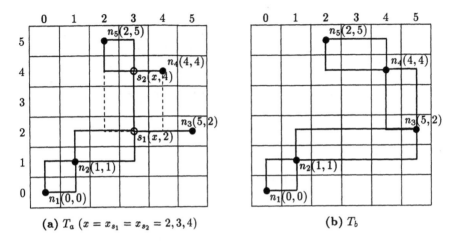

Figure 8.4. Edge-set trees

◇

If edge-set $E_{pq} \in T$, points p and q are adjacent. If p and q are adjacent, then p and q must be connected. However, a pair of connected points may not be a pair of adjacent points. If the number of points in an intersection of two point-sets, $|P_{pq} \cup P_{p'q'}|$, is equal to or greater than 1, edge-sets E_{pq} and $E_{p'q'}$ are adjacent. If $|P_{pq} \cup P_{p'q'}| \geq 1$, $E_{pq} \in T_1 \subset T$, and $E_{p'q'} \in T_2 \subset T$, then edge-set subtrees T_1 and T_2 are adjacent. If $|P_{pq} \cap P_{p'q'}| > 1$, then $P_{pq} \cap P_{p'q'}$ is called a redundant set of edge-sets E_{pq} and $E_{p'q'}$. There are three possible intersections of point-sets of two edge-sets: a point, a redundant set, and an empty set. The intersection of point-sets of two adjacent edge-sets is a point or a redundant set.

Example: In Figure 8.4a, n_1 and n_2 are adjacent, and n_1 and n_5 are connected, but not adjacent. In Figure 8.4b, $P_{n_2n_3} \cap P_{n_3n_4}$ is a redundant set of $E_{n_2n_3}$ and $E_{n_3n_4}$, because $|P_{n_2n_3} \cap P_{n_3n_4}| > 1$. Because of the redundant set, $l_{T_a} < l_{T_b}$. The intersection combinations of point-sets of two adjacent edge-sets are shown in Figure 8.5. Figure 8.5a shows the case of $|P_{pq} \cap P_{p'q'}| = 1$. Figures 8.5b to 8.5d show the intersection combinations of P_{pq} and $P_{p'q'}$ for the case of $|P_{pq} \cap P_{p'q'}| > 1$. Figure 8.5b shows the case of $P_{pq} \cap P_{p'q'} \in \{P_{pp'}, P_{pq'}, P_{p'q}, P_{qq'}\}$, Figure 8.5c shows the case of $\{p', q'\} \cap P_{pq} = p'$ or q', and Figure 8.5d shows the case of $\{p, q, p', q'\} \cap (P_{pq} \cap P_{p'q'}) = \emptyset$. The redundant set of $E_{n_2n_3}$ and $E_{n_3n_4}$ shown in Figure 8.4b belongs to the case shown in Figure 8.5c.

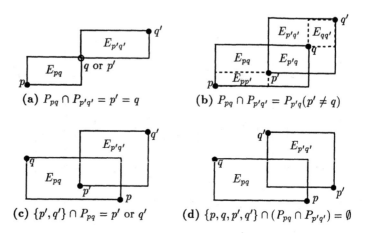

Figure 8.5. Possible intersections of point-sets of two adjacent edge-sets

◇

Theorem 8.2. If $|P_{pq} \cap P_{p'q'}| > 1$, then tree t with edges e_{pq} and $e_{p'q'}$ is not an MRST.

Proof: The intersection combinations of two adjacent edge-sets shown in Figure 8.5 are used in the proof. In Figure 8.5b, because $p' \neq q$, $|P_{pq} \cap P_{p'q'}| = |P_{p'q}| > 1$, and p, q, p' and q' can be connected by $\{E_{pp'}, E_{p'q}, E_{qq'}\}$, $l_{e_{pq}} + l_{e_{p'q'}} = l_{E_{pq}} + l_{E_{p'q'}} > l_{E_{pq}} + l_{E_{p'q'}} - l_{E_{p'q}} = l_{E_{pp'}} + l_{E_{p'q}} + l_{E_{qq'}}$. Hence, $l_{e_{pq}} + l_{e_{p'q'}}$ must not be the minimum length needed to connect p, q, p' and q'. Similarly, it can be proven that $l_{e_{pq}} + l_{e_{p'q'}}$ is not the minimum length of a subtree that connects p, q, p' and q' in the cases shown in Figure 8.4c and Figure 8.4d. Therefore, a tree having edges e_{pq} and $e_{p'q'}$ is not an MRST. □

In graph theory, a cycle is a graph with a set of edges $\{e_{p_ip_{i+1}}, e_{p_np_1} | i = 1, ..., n-1\}$. Similarly, $C_y = (E_{C_y}, V_{C_y})$, where $E_{C_y} = \{E_{p_ip_{i+1}}, E_{p_np_1} | i = 1, ..., n-1\}$ and $V_{C_y} = \{p_i | i = 1, ..., n\}$, is called an edge-set cycle. Its length is denoted as $l_{C_y} = \sum_{i=1}^{n-1} l_{E_{p_ip_{i+1}}} + l_{E_{p_np_1}}$. If $l_{E_{p_np_1}^{min}} < l_{E_{p_ip_{i+1}}^{max}}$ $(1 \leq i \leq n-1)$ for $P_{p_np_1} \cap P(T) = \{p_n, p_1\} \subset V_T$ and $E_{p_ip_{i+1}} \in E_{C_y} \subset (E_{p_np_1} \cup E_T)$, then removing $E_{p_ip_{i+1}}$ and reconnecting $\{E_{p_jp_{j+1}} | j = 1, ..., i-1\}$ and $\{E_{p_jp_{j+1}} | j = i+1, ..., n-1\}$ with a new edge-set, that connects p_n and p_1 with the length equal to $l_{E_{p_ip_{i+1}}^{max}}$, must generate a redundant set between the new edge-set and its adjacent edge-sets. The half-perimeter of the minimum rectangle enclosing the redundant set is called the redundant length of $E_{p_ip_{i+1}}$, and is measured by $l_{E_{p_ip_{i+1}}^{max}} - l_{E_{p_np_1}^{min}}$. Because T is converted to an edge-set tree having redundant sets, a set of trees contained by T is not a set of MRSTs.

Corollary 8.2.1. If $\{E_{pq}, E_{p'q'}\} \subset E_{C_y}$, $E_{pq} \in T$, $\{p', q'\} \subset V_T$, $E_{p'q'} \notin T$ and $l_{E_{pq}^{max}} > l_{E_{p'q'}^{min}}$, the total redundant length of edge-sets in T is at least equal to $l_{E_{pq}} - l_{E_{p'q'}}$.

Example: Figure 8.6 shows that an edge-set tree without an explicit redundant set is converted to an edge-set tree with a redundant set. The edge-set tree with $E_T = \{E_{p_i p_{i+1}} | i = 1, 2, 3\}$ shown in Figure 8.6a has no explicit redundant sets. Connecting p_1 and p_4, edge-set cycle C_y with $E_{C_y} \subset (E_{p_4 p_1} \cup E_T)$ is generated (see Figure 8.6b). Removing $E_{p_1 p_2}$ from T generates edge-set subtrees $T_1 = p_1$ and T_2 with $E_{T_2} = \{E_{p_2 p_3}, E_{p_3 p_4}\}$. Reconnecting T_1 and T_2 by $E_{p_1 p'_2} \notin T$ with $l_{E_{p_1 p'_2}} = l_{E_{p_1 p_2}}$ generates new edge-set tree T' with $E_{T'} = \{E_{p_1 p'_2}, E_{p_2 p_3}, E_{p_3 p_4}\}$ (Figure 8.6c). Because $l_{E_{p_1 p_2}} > l_{E_{p_1 p_4}}$ in cycle C_y, redundant set $P_{p_1 p'_2} \cap P_{p_3 p_4} = P_{p'_2 p_4}$ is generated. Therefore, the redundant length of $E_{p_1 p_2}$ is $l_{E_{p_1 p_2}} - l_{E_{p_4 p_1}} = l_{E_{p'_2 p_4}}$. Because $l_{T'} = l_T$ and $t' \in T'$ cannot be an MRST, $t \in T$ is also not an MRST. Redundant set $P_{p'_2 p_4}$ can be removed by replacing $E_{p_1 p'_2}$ with $E_{p_1 p_4}$.

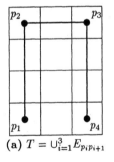
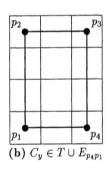
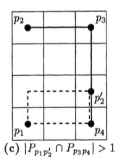

(a) $T = \cup_{i=1}^{3} E_{p_i p_{i+1}}$ (b) $C_y \in T \cup E_{p_4 p_1}$ (c) $|P_{p_1 p'_2} \cap P_{p_3 p_4}| > 1$

Figure 8.6. Transforming an edge-set tree by reconnecting an edge in an edge-cycle

◇

When $E_T = \{E_{p_i q_i} | i = 1, ..., n_T\}$ in which $|P_{p_i q_i} \cap P_{p_j q_j}| = 0$ or 1 ($i \neq j; i, j = 1, ..., n_T$), and no redundant length exists for $E_{p_i q_i}$ ($i = 1, ..., n_T$), edge-set tree T is called a no-redundancy edge-set tree (NR edge-set tree) and is denoted as T_{NR}. The edge-set tree T_a in Figure 8.4a is an NR edge-set tree. A set of trees that cannot be contained in an NR edge-set tree must not be a set of MRSTs.

8.2 Properties of Steiner Points

If $s \in V_s$ and $\{E_{p_is}|\ i = 1, ..., deg\,s\} \subset E_{T_{NR}}$, then

$$s = \bigcap_{i=1}^{deg\,s} P_{p_is} \qquad (8.7)$$

where $deg\,s$ is the **degree** of s, the number of edges that intersect at s. If $deg\,s = 1$, then $l_{E_{p_1s}}$ is a redundant length. Therefore, $deg\,s \neq 1$. Because s with $deg\,s = 2$ is a fake Steiner point, $deg\,s \neq 2$.

A selection function, $mid(a, b, c)$, selects the mid-value from three values. Rectangle R_{pq} is a minimum rectangle enclosing p_1, p_2 and p_3, if $x_p = min(x_{p_1}, x_{p_2}, x_{p_3})$, $x_q = max(x_{p_1}, x_{p_2}, x_{p_3})$, $y_p = min(y_{p_1}, y_{p_2}, y_{p_3})$, and $y_q = max(y_{p_1}, y_{p_2}, y_{p_3})$. The minimum length needed to connect three points is related to the minimum rectangle and the mid-value selection function.

Example: In Figure 8.7, $mid(x_{p_1}, x_{p_2}, x_{p_3}) = x_{p_2}$ and $mid(y_{p_1}, y_{p_2}, y_{p_3}) = y_{p_3}$, because $x_{p_1} < x_{p_2} < x_{p_3}$ and $y_{p_2} < y_{p_3} < y_{p_1}$. The coordinates of the minimum rectangle, R_{pq}, enclosing p_1, p_2 and p_3 are $x_p = min(x_{p_1}, x_{p_2}, x_{p_3}) = x_{p_1}$, $x_q = max(x_{p_1}, x_{p_2}, x_{p_3}) = x_{p_3}$, $y_p = min(y_{p_1}, y_{p_2}, y_{p_3}) = y_{p_2}$, and $y_q = max(y_{p_1}, y_{p_2}, y_{p_3}) = y_{p_1}$. Therefore, $p = (x_{p_1}, y_{p_2})$, $q = (x_{p_3}, y_{p_1})$, and R_{pq} is R_{ap_1} (designated by a dashed rectangle in Figure 8.7). Not every rectangle with its perimeter connecting three points is a minimum rectangle.

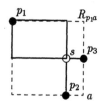

Figure 8.7. A Steiner point adjacent to three points

◇

Theorem 8.3. Point $s = (mid(x_{p_1}, x_{p_2}, x_{p_3}), mid(y_1, y_2, y_3))$ is the only Steiner point connecting points p_1, p_2 and p_3 with the minimum length, that equals a half-perimeter of the minimum rectangle enclosing p_1, p_2 and p_3, if and only if the vertices of a diagonal of the minimum rectangle are not two of p_1, p_2 and p_3.

Proof: Because four vertices of the minimum rectangle enclosing p_1, p_2 and p_3 are determined by two elements in $\{x_{p_1}, x_{p_2}, x_{p_3}\}$ and by two elements in $\{y_{p_1}, y_{p_2}, y_{p_3}\}$, p_1, p_2 or p_3 must be a vertex of the minimum rectangle.

Let p_1 be a vertex of the minimum rectangle R_{p_1a}, enclosing p_1, p_2 and p_3. Points p_2 and p_3 lie on two adjacent edges that are not adjacent to p_1 (Figure 8.7), or p_3 (p_2) lies on the opposite side of $e_{p_1p_2}$ ($e_{p_1p_3}$) in R_{p_1a}, if and only if $a \neq p_2$ and $a \neq p_3$.

In Figure 8.7, $p_1 \notin P_{p_2p_3}$ and $E_{p_1p_1} \natural E_{p_2p_3} = \emptyset$, because p_2 and p_3 lie on edges not adjacent to p_1. According to Corollary 8.1.2, $s = P_{p_1s} \cap P_{p_2p_3}$ is a vertex of $R_{p_2p_3}$ and the only point nearest to p_1. Moreover, E_{p_1s} is a bend edge-set. $p_3 = a$, if and only if $s = p_2$. However, it is impossible. Therefore, $s \notin \{p_2, p_3\}$. Because s is a vertex of $R_{p_2p_3}$, $x_s = x_{p_2}$ or x_{p_3}, $y_s = y_{p_2}$ or y_{p_3}, and $s = P_{p_1s} \cap P_{p_2p_3} = \cap_{i=1}^{3} P_{p_is}$. Further considering that s is not a vertex in the perimeter of R_{p_1a}, $x_s = mid(x_{p_1}, x_{p_2}, x_{p_3})$ and $y_s = mid(y_{p_1}, y_{p_2}, y_{p_3})$. Therefore, p_1, p_2 and p_3 are connected by $E_T = \{E_{p_is} | i = 1, 2, 3\}$ with a minimum total length equal to $l_T = \sum_{i=1}^{3} l_{E_{p_is}} = \frac{1}{2} l_{R_{p_1a}}$.

If p_3 lies on the opposite side of $e_{p_1p_2}$ in R_{p_1a}, then p_2 must be a vertex of R_{p_1a} and $E_{p_3p_3} \natural E_{p_1p_2} = \{p_3, s\}$, where $s \notin \{p_1, p_2\}$. According to Theorem 8.1, $s = \cap_{i=1}^{3} E_{p_is}$ is the only Steiner point connecting p_1, p_2 and p_3, with the minimum length $l_{E_{p_1p_2}} + l_{E_{p_3s}} = \frac{1}{2} l_{R_{p_1a}}$. $\square$

According to the proof of Theorem 8.3, if $E_{p_1p_1} \natural E_{p_2p_3} = \emptyset$, $s \notin \{p_2, p_3\}$ is a vertex of $R_{p_2p_3}$, E_{p_1s} is a bend edge-set, and E_{p_2s} and E_{p_3s} are two sides of $R_{p_2p_3}$.

Corollary 8.3.1. E_{p_2s} and E_{p_3s} must be straight edge-sets, if and only if E_{p_1s} is a bend edge-set in $\{E_{p_is} | s \in V_s(T_{NR}), i = 1, 2, 3\}$.

If $r \notin \cup_{i=1}^{3} P_{p_is}$, and E_{p_1s} is a bend edge-set, there must exist a redundant set $|P_{rs} \cap P_{p_is}| > 1$ ($1 \leq i \leq 3$). Furthermore, s is a fake Steiner point if $deg\, s = 2$.

Corollary 8.3.2. If $s \in V_s(T_{NR})$, and $E_{rs} \in E_{T_{NR}}$ is a bend edge-set, then s is an intersection of exactly three edge-sets.

If p_3 lies on the opposite side of $e_{p_1p_2}$ in the minimum rectangle enclosing p_1,

p_2 and p_3, E_{p_1s}, E_{p_2s} and E_{p_3s} are straight edge-sets. Only if $E_{p_3p_4}$ is a straight edge and $s \in P_{p_3p_4}$, then $\cap_{i=1}^4 P_{p_is} = s$.

Corollary 8.3.3. If $\{E_{p_is}|\ s \in V_s(T_{NR}), i = 1, ..., 4\}$, then E_{p_1s}, ..., and E_{p_4s} are all straight edge-sets.

If $\cap_{i=1}^4 P_{p_is} = s$ and $r \cap \cup_{i=1}^4 P_{p_1s} = \emptyset$, then E_{rs} is a bend edge-set. According to Corollary 8.3.2, *deg s* = 5 is impossible, and Corollary 8.3.4 can be proven.

Corollary 8.3.4. The degree of a fixed Steiner point is 3 or 4, and the degree of a given node is not more than 4 in an NR edge-set tree.

According to the proof of Theorem 8.3, $s = P_{p_3s} \cap P_{p_1p_2} = \cap_{i=1}^3 P_{p_is}$. $E_{p_1p_2}$ is thus broken into E_{p_1s} and E_{p_2s} in the NR edge-set tree connecting p_1, p_2 and p_3.

Corollary 8.3.5. If $s = P_{p_3s} \cap P_{p_1p_2} \in V_s(T_{NR})$, then T_{NR} with $E_{T_{NR}} = \{E_{p_is}|\ i = 1, 2, 3\}$ contains only one set of MRSTs for p_1, p_2 and p_3.

Corollary 8.3.5 also satisfies the case of $p_3 \in P_{p_1p_2}$. However, $s = p_3$ in this case. According to Corollary 8.3.5, connecting a point to an edge-set can be expressed by the following graph arithmetic for generating a T_{NR} for three points.

$$E_{p_1p_2} + p_3 = E_{p_1p_2} \bigsqcup E_{p_3s} \bigsqcup p_3 = E_{p_1p_2} \bigsqcup E_{p_3s} = \bigcup_{i=1}^3 E_{p_is} \qquad (8.8)$$

where operator "+" means "is connected with", and "$\sqcup$" is a union with an extra function for breaking $E_{p_1p_2}$ into $\{E_{p_1s}, E_{p_2s}\}$ by the Steiner point $s = (x_s, y_s)$, where $x_s = mid(x_{p_1}, x_{p_2}, x_{p_3})$ and $y_s = mid(y_{p_1}, y_{p_2}, y_{p_3})$.

Example: Figure 8.8 demonstrates the graph arithmetic to connect a point to an edge-set. The arithmetic includes: determining Steiner point s by Theorem 8.3, breaking $E_{p_1p_2}$ into $\{E_{p_1s}, E_{p_2s}\}$, and collecting E_{p_3s} to $\{E_{p_1s}, E_{p_2s}\}$. The function of "$\sqcup$" in $E_{p_1p_2} \sqcup E_{p_3s}$ includes breaking $E_{p_1p_2}$ by s and collecting E_{p_3s} to $\{E_{p_1s}, E_{p_2s}\}$. Similarly, the function of "$\sqcup$" in $E_{p_3s} \sqcup p_3$ includes breaking $E_{p_3p_3} = p_3$ by s. Because p_3 cannot be further broken into two edge-sets, the second "$\sqcup$" functions as a "$\cup$".

Figure 8.8. $E_{p_1 p_2} + p_3 = \cup_{i=1}^{3} E_{p_i s}$
◇

Implicitly expressing V_T by the subscripts of each edge-set, an edge-set tree, where no redundant set between any two adjacent edge-sets exists, can be expressed as a union of a set of edge-sets:

$$T = \bigcup_{i=1}^{n_T} E_{p_i q_i} \tag{8.9}$$

where each point in V_T satisfies Corollary 8.3.4 and a Steiner tree $t \in T$ can be expressed as $t = \bigcup_{i=1}^{n_T} e_{p_i q_i}$. If p_3 is removed from $T = \cup_{i=1}^{3} E_{p_i s}$, then $E_{p_3 s}$ is also removed, and s becomes a fake Steiner point, that can be eliminated by replacing $E_{p_1 s} \cup E_{p_2 s}$ with $E_{p_1 p_2}$. This graph arithmetic can be expressed as:

$$\cup_{i=1}^{3} E_{p_i s} - p_3 = \cup_{i=1}^{3} E_{p_i s} - \{E_{p_3 s}, p_3\} = E_{p_1 p_2} \tag{8.10}$$

Two NR edge-set subtrees can be generated by removing an edge-set from an NR edge-set tree. Equation 8.10 can be expressed as:

$$\cup_{i=1}^{3} E_{p_i s} - E_{p_3 s} = \{E_{p_1 p_2}, p_3\} \tag{8.11}$$

Example: Figures 8.9a and 8.9b demonstrate the graph arithmetics expressed by Equations 8.10 and 8.11. Whenever either p_3 or $E_{p_3 s}$ is removed, $E_{p_3 s}$ is removed, and $\cup_{i=1}^{2} E_{p_i s}$ is replaced by $E_{p_1 p_2}$ (see Corollary 8.1.5).

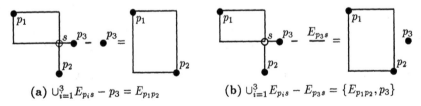

(a) $\cup_{i=1}^{3} E_{p_i s} - p_3 = E_{p_1 p_2}$ **(b)** $\cup_{i=1}^{3} E_{p_i s} - E_{p_3 s} = \{E_{p_1 p_2}, p_3\}$

Figure 8.9. Graph arithmetics for removing a node and an edge-set, respectively
◇

According to Theorem 8.3, at most 1 Steiner point exists when $|V_n| \leq 3$. According to Corollary 8.3.5, when a node is added to V_n, at most one new Steiner point can be generated.

Corollary 8.3.6. The maximum number of Steiner points in T_{NR} is $|V_n| - 2$.

A set of all edge-sets that always intersect at point $p \in V_{T_{NR}}$, even if p is moved, is called a set of dependent edge-sets of p, and is denoted as D_p. Similarly, a set of all edge-sets that is always adjacent to $E_{pq} \in E_{T_{NR}}$ is called a set of dependent edge-sets of E_{pq}, and is denoted as $D_{E_{pq}}$. Because $E_{pq} \in D_p$ and $E_{pq} \in D_q$,

$$D_{E_{pq}} = D_p \cup D_q - E_{pq} \qquad (8.12)$$

and the total number of edge-sets in $D_{E_{pq}}$ is

$$|D_{E_{pq}}| = deg\ p + deg\ q - 2. \qquad (8.13)$$

Example: In Figure 8.4a, $D_{n_2} = \{E_{n_1 n_2}, E_{n_2 s_1}\}$, $D_{s_1} = \{E_{n_2 s_1}, E_{n_3 s_1}, E_{s_1 s_2}\}$, $D_{E_{n_2 s_1}} = \{E_{n_1 n_2}, E_{n_3 s_1}, E_{s_1 s_2}\}$ and $D_{E_{s_1 s_2}} = \{E_{n_2 s_1}, E_{n_3 s_1}, E_{n_4 s_2}, E_{n_5 s_2}\}$. Therefore, $|D_{E_{n_2 s_1}}| = |D_{n_2}| + |D_{s_1}| - 2 = 2 + 3 - 2 = 3$.

However, not every edge is always connected to a moving Steiner point in an NR edge-set tree. As shown in Figure 8.10a, four edge-sets $E_{r_1 s}$, ..., $E_{r_4 s}$ intersect at Steiner point s, where $E_{r_4 s}$ is a parallel moving edge-set. When $E_{r_4 s}$ is moved up or down (Figure 10b), it will no longer intersect at s. Because $E_{r_4 s}$ is not always intersect at s, it is not a dependent edge-set of s. In this case, $|D_s| \neq 4$.

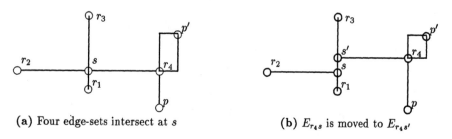

(a) Four edge-sets intersect at s (b) $E_{r_4 s}$ is moved to $E_{r_4 s'}$

Figure 8.10. There are three dependent edge-sets for a moving Steiner point
◇

Theorem 8.4. A dynamic Steiner point in an NR edge-set tree has exactly three dependent edge-sets.

Proof: According to Corollary 8.3.4, Steiner point $s \in V_s(T_{NR})$ can be an intersection of at most 4 straight edge-sets. If a bend edge-set is adjacent to s, according to Corollary 8.3.2, $|D_s| = 3$. If s is dynamic and $|D_s| = 4$, then all edge-sets that intersect at s are straight edge-sets, and s can be moved along each straight edge-set.

Assume $s = \cap_{i=1}^{4} P_{sr_i}$ (see Figure 8.10a), and s can be moved along $e_{r_1 r_3}$ or $e_{r_2 r_4}$. If s is moved along $e_{r_1 r_3}$ (see Figure 8.10b), then there must exist a parallel moving edge-set $E_{s' r_4}$, such that $E_{r_1 r_3} \natural E_{pp'} = \{ M_{s'}, M_{r_4} \mid y_{s'} = y_{r_4}, |M_s| = |M_{r_4}| > 1 \}$, where p and p' are nodes or Steiner points in specific locations and $E_{pp'}$ has been broken into $E_{pr_4} \cup E_{p'r_4}$. Here, $E_{s'r_4}$ is a parallel moving edge-set connecting $E_{ss'} \cup E_{s'r_3}$, and $E_{pr_4} \cup E_{p'r_4}$. Because of the movement of s', $E_{s'r_4}$ need not to always intersect at s. Therefore, $|D_s| = 4$ is impossible. $\square$

Theorem 8.5.   An edge-set in an NR edge-set tree has at most 4 dependent edge-sets.

Proof: According to Corollary 8.3.4 and Equation 8.13, the maximum $|D_{E_{pq}}|$ equals $deg\, p + deg\, q - 2 = 4 + 4 - 2 = 6$. From Corollary 8.3.3, if $|D_{E_{pq}}| = 5$, then E_{pq} must be a parallel moving edge-set. According to Theorem 8.4, $deg\, p = 4$ or $deg\, q = 4$ is impossible, because p and q are dynamic Steiner points. Therefore, the maximum $deg\, p = deg\, q = 3$ and $|D_{E_{pq}}| = 4$. $\square$

8.3   Cliques

A clique is an NR edge-set subtree that consists of moving edge-sets, and has:
- a minimum constant subtree length, independent of moving Steiner points;
- a minimum number of edge-sets, needed to maintain a maximum number of points in each domain.

A clique with n edge-sets is called an n-edge-set clique and is denoted as C_n. The length of clique C_n equals the total length of all edge-sets in C_n, and is denoted as l_{C_n}. A parallel moving edge-set is a clique, and is denoted as C_1, because its length is the minimum, and the domains of its vertices have been maximized. However, a one-vertex moving edge-set is not a clique, because its length is variable.

If T_{NR} consists of cliques and fixed edge-sets, all Steiner trees contained in T_{NR} have the same tree length, regardless of the movements of dynamic Steiner points. In contrast, if an edge-set tree has a moving edge-set that is not included in a clique, then there exists a redundant length, that equals the difference of the maximum and minimum lengths of the moving edge-set. Based on the minimum length of a clique and the maximum number of points in each domain, the domain of a dynamic Steiner point in an NR edge-set tree can be determined.

Theorem 8.6. The domain of a dynamic Steiner point in T_{NR} must be a set of points enclosed by a rectangle. (Proof omitted.)

Example: According to Theorem 8.4, a two-dimensional Steiner point, s, has three dependent edge-sets: E_{ps}, E_{rs}, and $E_{r's}$, where p is a fixed point (Figure 8.11). Because $x_s = x_r$, $y_s = y_{r'}$, $r \in e_{r_1 r_2}$, and $r' \in e_{r'_1 r'_2}$, M_s is dependent on $M_r = e_{r_1 r_2}$ and $M_{r'} = e_{r'_1 r'_2}$, and is a point-set enclosed by a rectangle. Two adjacent edges of this rectangle are parallel to $e_{r_1 r_2}$ and $e_{r'_1 r'_2}$, and have the lengths equal to $l_{e_{r_1 r_2}}$ and $l_{e_{r'_1 r'_2}}$, respectively. If s is moved within this rectangle, the total length needed to connect p, r and r', $l_{ps} + l_{rs} + L_{r's}$, is constant.

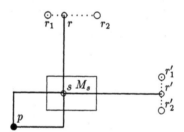

Figure 8.11. Domains of a dynamic Steiner points

◇

According to Theorems 8.1 and 8.6, $l_{E_{ps}} + l_{E_{rs}} + l_{E_{r's}}$ is the minimum length needed to connect p, s, r, and r', if $r \in e_{r_1 r_2}$, $r' \in e_{r'_1 r'_2}$, $l_{E_{rs}}^{min} \geq l_{e_{r_1 r_2}} = |M_r|$, and $l_{E_{r's}}^{min} \geq l_{e_{r'_1 r'_2}} = |M_{r'}|$ (see Figure 8.11). If one of the three edge-sets is removed, the length of other two edge-sets is not constant, unless s is reduced to a one-dimensional Steiner point. Edge-set subtree $E_{ps} \cup E_{rs} \cup E_{r's}$ satisfies the condition of a clique, and therefore is a 3-edge-set clique (C_3).

In $C_3 = E_{s_p} \cup E_{s_r} \cup E_{s_{r'}}$ (see Figure 8.11), r and r' are called dynamic end-vertices, p is called a fixed end-vertex, as well as M_r and $M_{r'}$ are two sets of points lying on two edges perpendicular to E_{sr} and $E_{sr'}$, respectively. Therefore, C_3 can be described by Corollary 8.6.1, and the dimension of a dynamic Steiner point can be determined by Corollaries 8.6.2 and 8.6.3.

Corollary 8.6.1. If a two-dimensional Steiner point is adjacent to a fixed point, then the dependent edge-sets of the Steiner point and all their end-vertices form a 3-edge-set clique (C_3).

Corollary 8.6.2. If $C_3 = E_{ps} \cup E_{rs} \cup E_{r's}$, where E_{ps} is a bend edge-set, then M_r and $M_{r'}$ must be two sets of points lying on two straight edges perpendicular to straight edge-sets E_{rs} and $E_{r's}$, respectively.

Corollary 8.6.3. If r_1 and r_2 are dynamic end-vertices of C_3, with their domains $M_r = P_{r_1 r_2}$ $(x_{r_1} < x_{r_2}, \ y_{r_1} = y_{r_2})$ and $M_{r'} = P_{r'_1 r'_2}$ $(x_{r'_1} = x_{r'_2}, \ y_{r'_1} < y_{r'_2})$, respectively, then $M_s = P_{ab}$ where $a = (x_{r_1}, y_{r'_1})$ and $b = (x_{r_2}, y_{r'_2})$.

A rectangle enclosing a clique with the minimum perimeter is called the minimum rectangle enclosing that clique. The minimum rectangle enclosing a 3-edge-set clique is the minimum rectangle enclosing three end-vertices of the clique. According to Corollaries 8.6.1 to 8.6.3 and Theorem 8.3, Corollary 8.6.4 can be obtained.

Corollary 8.6.4. If R is the minimum rectangle enclosing C_3, then $l_{C_3} = \frac{1}{2} l_R$.

If p is also a dynamic Steiner point in Figure 8.11, according to Corollary 8.6.2, a 5-edge-set clique (C_5), which includes a two-vertex moving edge-set and its dependent edge-sets, as well as four dynamic end-vertices lying on the perimeter of a rectangle, can be explained by:

Corollary 8.6.5. Dependent edge-sets of a pair of adjacent two-dimensional Steiner points and their end-vertices form a 5-edge-set clique (C_5), whose length equals a half-perimeter of the minimum rectangle enclosing this clique.

Similar to the domain of the two-dimensional Steiner points in a 3-edge-set clique, domains of two-dimensional Steiner points in a 5-edge-set clique depend on the domains of their adjacent dynamic end-vertices. However, movements of two-dimensional Steiner points may or may not depend upon each other. According to Corollaries 8.6.4 and 8.6.5, the relationship of a pair of two-dimensional

Steiner points in a 5-edge-set clique can be expressed by Corollary 8.6.6. Further-more, the domain of a moving end-vertex of a 3-edge-set or 5-edge-set clique must also satisfy Corollary 8.6.7.

Corollary 8.6.6. If $C_5 = E_{sr_1} \cup E_{sr_2} \cup E_{ss'} \cup E_{s'r_1'} \cup E_{s'r_2'}$, then dynamic Steiner points s and s' must satisfy one of the following conditions:

(a) If $x_{r_1} < x_s = x_{r_2}$ and $y_{r_2} < y_s = y_{r_1}$, then $x_s \leq x_{s'} = x_{r_2'} < x_{r_1'}$ and
   $y_s \leq y_{s'} = y_{r_1'} < y_{r_2'}$.

(b) If $x_{r_1} < x_s = x_{r_2}$ and $y_{r_1} = y_s < y_{r_2}$, then $x_s \leq x_{s'} = x_{r_2}' < x_{r_1'}$ and
   $y_{r_2}' < y_{s'} = y_{r_1'} \leq y_s$.

Example: Figures 8.12a and 8.12b show two 5-edge-set cliques, in which two-dimensional Steiner points satisfy conditions (a) and (b) of Corollary 8.6.6, re-spectively. If s' is fixed in Figure 8.12a, then a 3-edge-set clique can be obtained (see Figure 12c). According to Corollary 8.6.6, s cannot be moved to (x,y) in Figure 8.12a, where $x \leq x_{s'}$ or $y \leq y_{s'}$. The domain of each dynamic end-vertex of a clique is a set of points lying on the perimeter of the minimum rectangle. Figure 8.12b shows another clique, that is enclosed by the same rectangle.

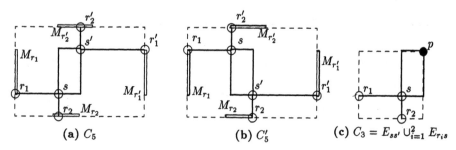

(a) C_5 (b) C_5' (c) $C_3 = E_{ss'} \cup_{i=1}^2 E_{r_is}$

Figure 8.12. 5-edge-set cliques and a 3-edge-set clique
$$(C_5 = \cup_{i=1}^2 E_{r_is} \cup E_{ss'} \cup_{i=1}^2 E_{s'r_i'}$$
$$C_5' = E_{r_1s} \cup E_{r_2's} \cup E_{ss'} \cup E_{s'r_1'} \cup E_{s'r_2})$$

◇

Corollary 8.6.7. If r and E_{rs} are a dynamic end-vertex and an edge-set of a 3-edge-set clique or a 5-edge-set clique, respectively, then $|M_r| \leq l_{E_{rs}^{min}}$.

Figure 8.12 also shows that two 5-edge-set cliques may exists in the same minimum rectangle.

Theorem 8.7. There exist two different 5-edge-set cliques with $M_{r_1} \in e_{aa'}$, $M_{r_2} \in e_{bb'}$, $M_{r_1'} \in e_{cc'}$ and $M_{r_2'} \in e_{dd'}$ as the domains of end-vertices, which in turn lie on 4 sides of a rectangle, if and only if there exist parallel moving edge-sets E_{pq} and E_{fg}, such that $e_{aa'} \natural e_{cc'} = \{M_p, M_q | \ y_p = y_q, |M_p| = |M_q| \geq 1\}$, $e_{bb'} \natural e_{dd'} = \{M_f, M_g | \ x_f = x_g, |M_f| = |M_g| \geq 1\}$, $|M_f| \neq 1$ if $|M_p| = |M_f|$, $l_{e_{aa'}} \leq min(|x_p - x_b|, |x_p - x_{b'}|, |x_p - x_d|, |x_p - x_{d'}|)$, $l_{e_{bb'}} \leq min(|y_f - y_a|, |y_f - y_{a'}|, |y_f - y_c|, |y_f - y_{c'}|)$, $l_{e_{cc'}} \leq min(|x_q - x_b|, |x_q - x_{b'}|, |x_q - x_d|, |x_q - x_{d'}|)$, and $l_{e_{dd'}} \leq min(|y_g - y_a|, |y_g - y_{a'}|, |y_g - y_c|, |y_g - y_{c'}|)$. (Proof omitted.)

Example: If parallel moving edge-sets exist between edge-sets with point-sets M_{r_1} and $M_{r_1'}$ and between edge-sets with point-sets M_{r_2} and $M_{r_2'}$ in Figure 8.12a, then r_1, r_1', r_2 and r_2' can be moved to $(x_{r_1}, y_{r_1'})$, $(x_{r_1'}, y_{r_1})$, $(x_{r_2'}, y_{r_2})$ and $(x_{r_2}, y_{r_2'})$, respectively, and the new clique shown in Figure 8.12b is obtained.

◇

Figures 8.12a and 8.12b also show that $|M_s \cap M_{s'}| > 1$ if two 5-edge-set cliques exist in the same minimum rectangle. Therefore, Theorem 8.7 can be further described as Corollary 8.7.1, and the relationship of end-vertices of the two cliques can be described by Corollary 8.7.2. Corollary 8.7.3 can be obtained from Theorems 8.3 and 8.7, because the domain of the two-dimensional Steiner point in a 3-edge-set tree is only determined by the domains of two end-vertices.

Corollary 8.7.1. Two 5-edge-set cliques enclosed by the same minimum rectangle exist, if and only if there is more than one point in the intersection of the domains of two-dimensional Steiner points in a 5-edge-set clique.

Corollary 8.7.2. If end-vertices $r \in C_5$ and $r' \in C_5'$ lie on the same side of the minimum rectangle enclosing C_5 and C_5', then $M_r \cap M_{r'} \neq \emptyset$.

Corollary 8.7.3. There is only one C_3 within a minimum rectangle.

No clique can independently exist in an NR edge-set tree without connecting to other cliques, because of dynamic end-vertices. The structures of interfaces among nodes and cliques, stated in Corollaries 8.6.1 and 8.6.5, are described by Theorem 8.8 and its Corollaries.

Theorem 8.8.   If r and r' are adjacent dynamic end-vertices of two cliques enclosed by different minimum rectangles, respectively, then

(a) Two nodes are adjacent to r and r', respectively, if $E_{rr'}$ is a bend edge-set;

(b) Two nodes terminate a straight line crossing r and r', if $E_{rr'}$ is a straight edge-set.   (Proof omitted.)

Example: According to Theorem 8.8, the interfaces among cliques and given nodes are shown in Figure 8.13. The structures of interfaces described by the conclusion (a) of Theorem 8.8 are shown in Figure 8.13a, where r and r' are dynamic end-vertices of C_i and C_j, respectively, and p and q are given nodes. Figure 8.13b shows a special case of the structures in Figure 8.13a. The structure of an interface described by the conclusion (b) of Theorem 8.8 is shown in Figure 8.13c, in which p and q are nodes, and r, r', ..., and $r^{(n)}$ are dynamic end-vertices of n cliques enclosed by n different minimum rectangles. Note that one of the dependent edge-sets of each dynamic end-vertex is degenerated in every interface.

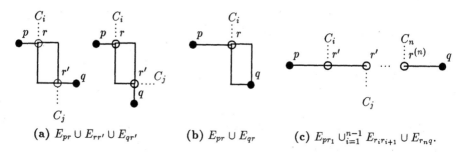

(a) $E_{pr} \cup E_{rr'} \cup E_{qr'}$ (b) $E_{pr} \cup E_{qr}$ (c) $E_{pr_1} \cup_{i=1}^{n-1} E_{r_i r_{i+1}} \cup E_{r_n q}$.

Figure 8.13.   Interfaces among cliques and given nodes

◇

Each interface in Figure 8.13 has a constant length and satisfies the conditions of a clique. Because an edge adjacent to a Steiner point is degenerated, cliques in Figure 8.13 are called degenerated cliques, and the Steiner point with one dependent edge-set degenerated is called a degenerated Steiner point. To distinguish cliques C_1 (parallel moving edge-set), C_3 and C_5 from the degenerated cliques, C_1, C_3 and C_5 are also called non-degenerated cliques. A degenerated clique with n edge-sets is denoted as C_{nd}. The types of degenerated cliques are described by Corollaries 8.8.1 to 8.8.4.

Corollary 8.8.1. A degenerated clique is enclosed by the minimum rectangle with two nodes as vertices, and its length equals a half-perimeter of the minimum rectangle.

Corollary 8.8.2. A degenerated clique that includes a bend edge-set with two degenerated Steiner points as vertices is a 3-edge-set degenerated clique (C_{3d}).

Corollary 8.8.3. A degenerated clique with only one degenerated Steiner point is a 2-edge-set degenerated clique (C_{2d}).

Corollary 8.8.4. A degenerated clique that includes a straight edge-set with two degenerated Steiner points as vertices is an n-edge-set degenerated clique with n straight edge-sets on the same line.

Example: Each degenerated clique in Figure 8.13 is enclosed by a minimum rectangle. Note that the minimum rectangle enclosing the cliques in Figure 8.13c is a straight edge (a special rectangle with $y_p = y_q$). Figure 8.13a shows that the degenerated clique with a bend edge-set, $E_{rr'}$, is a 3-edge-set degenerated clique. Figure 8.13b shows that a degenerated clique with only one degenerated Steiner point is a 2-edge-set degenerated clique. The degenerated clique in Figure 8.13c is also a 2-edge-set degenerated clique if $n = 1$. Figure 8.13c shows that all edge-sets are straight, if an edge-set with two degenerated points as vertices is straight.
◇

Because each two-dimensional Steiner point is included in a 3-edge-set clique or a 5-edge-set clique, each one-dimensional Steiner point is an end-vertex of a non-degenerated clique, each non-degenerated clique is adjacent to a degenerated clique, and each degenerated clique is adjacent to given nodes, each moving edge-set in an NR edge-set tree is an edge-set of a clique, and cliques C_1, C_3, C_5, C_{2d}, C_{3d} and C_{nd} $(n > 3)$ are all different types of cliques, that form an NR edge-set tree. Therefore, T_{NR} consists of fixed edge-sets and cliques:

$$T_{NR} = \bigcup_{n=1,3,5} \left(\bigcup_{i_n=0}^{n_{c_n}-1} C_n^{(i_n)} \right) \bigcup \bigcup_{l=2}^{n_d} \left(\bigcup_{j_l=0}^{n_{dl}-1} C_{ld}^{(j_l)} \right) \bigcup \left(\bigcup_{k=1}^{n_E} E_{p_k q_k} \right) \tag{8.14}$$

where n_{c_n} is the number of non-degenerated cliques with n edge-sets, n_d is the maximum number of edge-sets in a degenerated clique, n_{dl} is the number of degenerated cliques with l edge-sets, n_E is the number of fixed edge-sets, $C_n^{(i)}$ is the ith non-degenerated clique with n edge-sets, and $C_{ld}^{(j)}$ is jth degenerated clique with l edge-sets.

Example: The NR edge-set tree in Figure 8.14 can be expressed as $T_{NR} = C_1 \cup C_5 \cup (\cup_{i=0}^3 C_{2d}^{(i)}) \cup C_{3d}$, where $C_1 = E_{s_1 s_2}$, $C_5 = E_{s_3 s_4} \cup E_{s_8 s_4} \cup E_{s_4 s_5} \cup E_{s_5 s_6} \cup E_{s_5 s_7}$, $C_{2d} = E_{n_1 s_1} \cup E_{n_2 s_1}$, $C_{2d}' = F_{n_4 s_6} \cup E_{n_5 s_6}$, $C_{2d}'' = E_{n_6 s_7} \cup E_{n_7 s_7}$, $C_{2d}''' = E_{n_8 s_8} \cup E_{n_9 s_8}$, and $C_{3d} = E_{n_3 s_3} \cup E_{s_2 s_3} \cup E_{n_{10} s_2}$. In this NR edge-set tree, s_4 and s_5 are two-dimensional Steiner points, and s_1, s_2, s_3, s_6, s_7, and s_8 are one-dimensional Steiner points, where $x_{n_1} \le x_{s_1} = x_{s_2} \le x_{n_2}$, $x_{s_2} \le x_{s_3} = x_{s_4} \le x_{n_3}$, $max(x_{n_7}, x_{s_4}) \le x_{s_5} = s_7 \le x_{n_6}$, $y_{n_5} \le y_{s_5} = y_{s_6} \le y_{n_4}$, and $y_{n_8} \le y_{s_4} = y_{s_8} \le y_{n_9}$. Note that each straight moving edge-set in every non-degenerated clique can be parallelly moved.

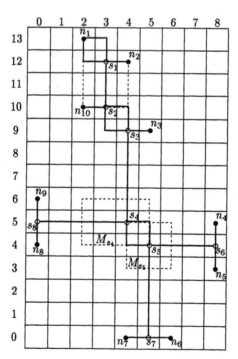

Figure 8.14. An NR edge-set tree for 10 nodes

8.4 Exchangeable Edge-Sets and Trees

Applying the graph arithmetic expressed by Equation 8.11, separate NR edge-set subtrees T_{NR_1}, ... and $T_{NR_{n_{ev}}}$, for n_{ev} sub-sets of nodes, can be generated by removing a non-degenerated clique with n_{ev} end-vertices from T_{NR}, expressed by Equation 8.14. The graph arithmetic for removing a non-degenerated clique from an NR edge-set tree can be expressed as:

$$T_{NR} - C_n = \{T_{NR_1}, \ldots, T_{NR_{n_{ev}}}\} \tag{8.15}$$

where $n_{ev} = 2$ for C_1, $n_{ev} = 3$ for C_3, $n_{ev} = 4$ for C_5, $P(T_{NR_i}) \cap P(T_{NR_j}) = \emptyset$ ($i \neq j; 1 \leq i, j \leq n_{ev}$), and $l_{T_{NR}} = \sum_{i=1}^{n_{ev}} l_{T_{NR_i}} + l_{C_n}$. Because l_{C_n} is a constant, $l_{T_{NR_i}}$ ($i = 1, ..., n_{ev}$) is also a constant. A degenerated clique, an interface among non-degenerated cliques, is determined by the non-degenerated cliques in an NR edge-set tree. Therefore, the removal of a degenerated clique from an NR edge-set tree will not be discussed.

Applying the graph arithmetic expressed by Equation 8.8, T_{NR_1}, ... and $T_{NR_{n_{ev}}}$ can be reconnected as an NR edge-set tree by a clique with the same length and the same number of end-vertices as those of C_n. The graph arithmetic for re-connecting $\{T_{NR_i} | i = 1, ..., n_{ev}\}$ with a non-degenerated clique to form an NR edge-set tree can be expressed as:

$$T'_{NR} = \bigcup_{i=1}^{n_{ev}} T_{NR_i} \bigsqcup C'_n \tag{8.16}$$

If $C'_n = C_n$, then $T'_{NR} = T_{NR}$. Otherwise, a new NR edge-set tree is generated. Therefore, more than one NR edge-set trees exists for the same set of nodes.

Example: Figures 8.15a and 8.15b show the results obtained by removing $C_1 = E_{s_1 s_2}$ and $C_5 = E_{s_3 s_4} \cup E_{s_8 s_4} \cup E_{s_4 s_5} \cup E_{s_5 s_6} \cup E_{s_5 s_7}$ from T_{NR} in Figure 8.14. According to the graph arithmetic expressed by Equation 8.11, after removing C_1 from T_{NR}, $C_{2d} = E_{n_1 s_1} \cup E_{n_2 s_1} \in T_{NR}$ is replaced with fixed edge-set $E_{n_1 n_2}$, and $C_{3d} = E_{n_3 s_3} \cup E_{s_2 s_3} \cup E_{n_{10} s_2} \in T_{NR}$ is replaced with 2-edge-set degenerated clique $E_{n_3 s_3} \cup E_{n_{10} s_3}$ (see Figure 8.15a). Figure 8.15b shows $T_{NR} - C_5$, in which $C_{3d} \in T_{NR}$, $C'_{2d} = E_{n_4 s_6} \cup E_{n_5 s_6} \in T_{NR}$, $C''_{2d} = E_{n_6 s_7} \cup E_{n_7 s_7} \in T_{NR}$, and $C'''_{2d} = E_{n_8 s_8} \cup E_{n_9 s_8} \in T_{NR}$ are replaced with 2-edge-set clique $E_{n_{10} s_2} \cup E_{s_2 n_3}$, $E_{n_4 n_5}$, $E_{n_6 n_7}$, and $E_{n_8 n_9}$, respectively.

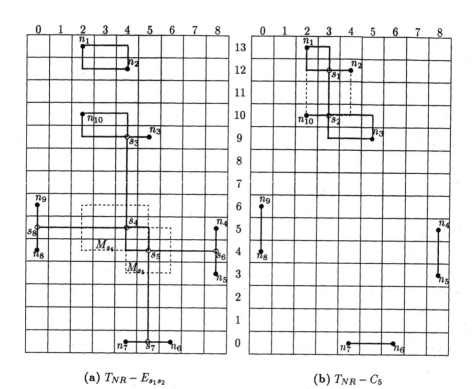

(a) $T_{NR} - E_{s_1 s_2}$ (b) $T_{NR} - C_5$

Figure 8.15. Removing cliques from an NR edge-set tree

Let $E_{pq} \in T_{NR}$ be a fixed edge-set or a moving edge-set. Applying Equation 8.11, Equation 8.14 can be written as:

$$T_{NR} = T_{NR_1} \bigsqcup E_{pq} \bigsqcup T_{NR_2} \tag{8.17}$$

where $P(T_{NR_1}) \cap P(T_{NR_2}) = \emptyset$, and

$$l_{T_{NR}} = l_{T_{NR_1}} + l_{T_{NR_2}} + l_{E_{pq}} \tag{8.18}$$

Although $l_{T_{NR}}$ is a constant, $l_{T_{NR_1}}$ and $l_{T_{NR_2}}$ are not constants if E_{pq} is a moving edge-set. Only if $l_{E_{pq}}$ is maximized, $l_{T_{NR_1}} + l_{T_{NR_2}}$ is minimized, and no redundant length exists in T_{NR_1} and T_{NR_2}. Let $T_{NR}|_{(E_{pq}^{max} \to E_{pq})}$ be an edge-set tree obtained by replacing E_{pq} with E_{pq}^{max} in T_{NR}, where "$\to$" implies "is assigned to " or "replaces". Then

$$T_{NR}|_{(E_{pq}^{max} \to E_{pq})} = T_{NR_1} \bigsqcup E_{pq}^{max} \bigsqcup T_{NR_2} \tag{8.19}$$

and

$$T_{NR} - E_{pq} = T_{NR}|_{(E_{pq}^{max} \to E_{pq})} - E_{pq}^{max} = \{T_{NR_1}, T_{NR_2}\} \qquad (8.20)$$

Example: Figure 8.16 shows the graph arithmetic for removing $E_{s_4 s_5}$ from T_{NR} in Figure 8.14, where $2 \le x_{s_4} \le max(5, x_{s_5})$, $max(4, y_{s_5}) \le y_{s_4} \le 6$, $max(x_{s_4}, 4) \le x_{s_5} \le 6$, $3 \le y_{s_5} \le min(y_{s_4}, 5)$, and $0 \le l_{E_{s_4 s_5}} \le 7$. When $s_4 = (2, 6)$ and $s_5 = (6, 3)$, $E_{s_4 s_5}^{max} \to E_{s_4 s_5}$. $T_{NR}|_{(E_{s_4 s_5}^{max} \to E_{s_4 s_5})}$ means that $E_{s_4 s_5}$ is replaced by $E_{s_4 s_5}^{max}$, or $E_{s_4 s_5}^{max}$ is assigned to $E_{s_4 s_5}$ (see Figure 8.16a). Figure 8.16b shows two NR edge-set subtrees, obtained by removing $E_{s_4 s_5}$ from T_{NR}: $T_{NR} - E_{s_4 s_5}$.

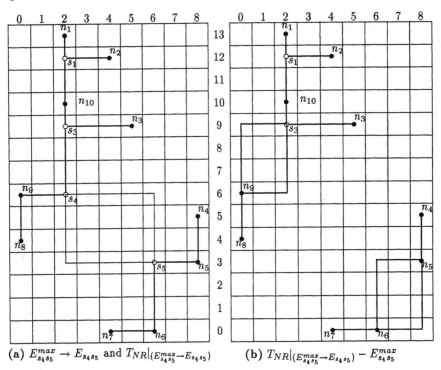

(a) $E_{s_4 s_5}^{max} \to E_{s_4 s_5}$ and $T_{NR}|_{(E_{s_4 s_5}^{max} \to E_{s_4 s_5})}$ (b) $T_{NR}|_{(E_{s_4 s_5}^{max} \to E_{s_4 s_5})} - E_{s_4 s_5}^{max}$

Figure 8.16. Removing a moving edge-set from an NR edge-set tree

◇

Equations 8.19 and 8.20 also satisfy the edge-set tree expressed by Equation 8.5. If $\{T_1, T_2\} = T - E_{p_i q_i}$, where $P(T_1) \cap P(T_2) \ne \emptyset$, then $E_{p_i q_i}$ is a redundant edge-set. Theorem 8.9 can be proven by using Equation 8.20.

Theorem 8.9. An edge-set tree without a redundant length is also an edge-set tree without redundant sets.

Proof: Assuming $|P_{p_1 q_1} \cap P_{p_2 q_2}| > 1$ in edge-set tree T, where no redundant length exists for any edge-set. If $\{T_1, T_2\} = T - E_{p_1 q_1}$, and $E_{p'q'} \notin T$, where $p' = P_{p'q'} \cap P(T_1)$ and $q' = P_{p'q'} \cap P(T_2)$, then there exists a cycle $C_y \supset \{E_{p_1 q_1}, E_{p'q'}\}$, where $l_{E_{p_1 q_1}} \leq l_{E_{p'q'}}$. If $p' = p_1 \in P_{p_1 q_1} \cap P(T_1)$, $E_{p_2 q_2} \in T_2$, and $P_{p'q'} \cap P_{p_2 q_2} = q'$, then $l_{E_{p'q'}} < l_{E_{p_1 q_1}}$. Therefore, $|P_{p_1 q_1} \cap P_{p_2 q_2}| > 1$ is impossible. $\square$

For every point $p \in P(T_{NR_1})$ and every point $q \in P(T_{NR_2})$, the minimum distance between p and q is called the distance of T_{NR_1} and T_{NR_2}. Similarly, for every point $p \in P_{p'q'} \subset P(T_{NR_1})$ and every point $q \in P_{p''q''} \subset P(T_{NR_2})$, the minimum distance between p and q is called the distance of edge-sets $E_{p'q'}$ and $E_{p''q''}$. If $l_{E_{p'q'}}$ and $l_{E_{p''q''}}$ are not constant, the distance between $E_{p'q'}$ and $E_{p''q''}$ can be found by comparing $E_{p'q'}^{max}$ and $E_{p''q''}^{max}$. If $E_{p'q'}$ or $E_{p''q''}$ is a parallel moving edge-set, the distance between them can be obtained by moving $E_{p'q'}$ and $E_{p''q''}$ parallelly. The distance between two edge-set subtrees can be obtained by comparing the distance between edge-sets of different edge-set subtrees. Let the distance between $E_{p'q'}^{max} \in T_{NR_1}$ and $E_{p''q''}^{max} \in T_{NR_2}$ be the distance of T_{NR_1} and T_{NR_2}. If $P_{p'q'} \cap P_{ss'} = s$ and $P_{p''q''} \cap P_{ss'} = s'$, then

$$T = (T_{NR_1} - E_{p'q'}) \bigsqcup (E_{p'q'}^{max} \bigsqcup E_{ss'}^{min} \bigsqcup E_{p''q''}^{max}) \bigsqcup (T_{NR_2} - E_{p''q''}). \qquad (8.21)$$

If s and s' are Steiner points, M_s and $M_{s'}$ can be determined by Corollary 8.6.6 and $E_{ss'}^{min}$ can be recovered to $E_{ss'}$. Denoting $M_s \to s$ and $M_{s'} \to s'$ as operations for expanding s to M_s and s' to $M_{s'}$, respectively, the reconnected NR edge-set tree can be expressed as:

$$
\begin{aligned}
T'_{NR} &= T|_{M_s \to s, \, M_{s'} \to s'} \\
&= [(T_{NR_1} - E_{p'q'}) \bigsqcup E_{p'q'}] \bigsqcup E_{ss'} \bigsqcup [(T_{NR_2} - E_{p''q''}) \bigsqcup E_{p''q''}] \\
&= T_{NR_1} \bigsqcup E_{ss'} \bigsqcup T_{NR_2} \\
&= (T_{NR} - E_{rr'}) \bigsqcup E_{ss'} \qquad\qquad\qquad\qquad\qquad (8.22)
\end{aligned}
$$

where $T|_{M_s \to s, \, M'_s \to s'}$ is an edge-set tree in which the domains of s and s' are recovered to M_s and $M_{s'}$, respectively, and $T'_{NR} = T_{NR_1} \sqcup E_{rr'} \sqcup T_{NR_2}$. Equation 8.22 shows that a new NR edge-set tree can be generated by replacing an edge-set in an NR edge-set tree by an edge-set with the same minimum length. If $E_{rr'} = E_{ss'}$ in Equation 8.22, then $T'_{NR} = T_{NR}$. Equations 8.20 and 8.22 also show that an NR edge-set tree can be iteratively generated from $T = \cup_{i=1}^{n_T} E_{p_i q_i}$, which is not an NR edge-set tree, by replacing $E_{p_i q_i}$ with $E_{p'_i q'_i}$ under the conditions:

(a) $T = T - E_{p_i q_i}$, if $T - E_{p_i q_i}$ is a connected graph;

(b) $T = (T - E_{p_i q_i}) \sqcup E_{p'_i q'_i}$, if $T - E_{p_i q_i} = \{T_1, T_2\}$, where $P(T_1) \cap P(T_2) = \emptyset$,
$E_{p'_i q'_i} \cap P(T_1) = p'_i$, $E_{p'_i q'_i} \cap P(T_2) = q'_i$, and $l_{E_{p'_i q'_i}} \le l_{E_{p_i q_i}}$.

Example: Figure 8.17 shows an operation for connecting NR edge-set subtrees. In Figure 8.17a, $T_{NR_1} = C_{2d} \cup C_1 \cup C'_{2d} = (E_{p_1 s_1} \cup E_{q_1 s_1}) \cup E_{s_1 s_2} \cup (E_{p_2 s_2} \cup E_{q_2 s_2})$ consists of a parallel moving edge-set (C_1) and two adjacent degenerated cliques $(C_{2d}$ and $C'_{2d})$. To determine the distance between two separate edge-sets, $E_{q_2 s_2}$ and $E_{pp} = p$, $E_{q_2 s_2}$ is replaced by $E_{q_2 s_2}^{max}$, or $E_{q_2 s_2}^{max}$ is assigned to $E_{q_2 s_2}$: $E_{q_2 s_2}^{max} \rightarrow E_{p_2 q_2}$ (Figure 8.17b). From $E_{p_2 q_2}$ and p, the distance between two NR edge-set subtrees can be determined. In the operation $E_{p_2 q_2} \sqcup E_{p s_3}$, $E_{p_2 q_2}$ is broken into $E_{p s_3} \cup E_{q_2 s_3}$. Because $E_{q_2 q_3} \sqcup E_{p s_3} = \{M_{s_3}, M_{s_4} \mid x_3 = x_4, |M_{s_3}| = |M_{s_4}| = 2\}$, $E_{s_3 s_4}$ is a parallel moving edge-set (Figure 8.17c).

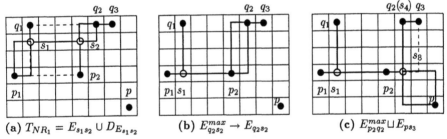

(a) $T_{NR_1} = E_{s_1 s_2} \cup D_{E_{s_1 s_2}}$ (b) $E_{q_2 s_2}^{max} \rightarrow E_{q_2 s_2}$ (c) $E_{p_2 q_2}^{max} \sqcup E_{p s_3}$

Figure 8.17. Operations of $[(T_{NR_1} - E_{p_2 q_2}) \cup E_{p_2 q_2}^{max}] \sqcup E_{p s_3} \cup p$

⬦

If $T'_{NR} - E_{p'q'} = T_{NR} - E_{pq}$ and $l_{E_{p'q'}} = l_{E_{pq}}$, then $E_{p'q'}$ is called an exchangeable edge-set of E_{pq}, and T'_{NR} is an exchangeable edge-set tree of T_{NR}. Similarly, if $T'_{NR} - C' = T_{NR} - C$ and $l_{C'} = l_C$, then C' is called an exchangeable clique of C. The exchangeable edge-sets of a fixed edge-set or of a parallel moving edge-set can be found by applying Equations 8.19 to 8.22.

According to Corollary 8.7.3, no exchangeable clique exists for C_3. However, according to Theorem 8.7, an exchangeable clique may exists for C_5. Even if $C_5 = E_{s_1 s_2} \cup D_{E_{s_1 s_2}}$ can be exchanged with $C'_5 = E_{s'_1 s'_2} \cup D_{E_{s'_1 s'_2}}$, $E_{s_1 s_2}$ and $E_{s'_1 s'_2}$ are not exchangeable, because $C_5 - E_{s_1 s_2} \ne C'_5 - E_{s'_1 s'_2}$. Theorem 8.10 explains the relationship between exchanging C_5 with C'_5 and exchanging $E_{rs} \in C_5$ with $E_{r's'} \in C'_5$.

Theorem 8.10. If $E_{rs} \in C_5 \subset T_{NR}$, and $E_{r's'} \in C_5'$, then $C_5' = (C_5 - E_{rs}) \sqcup E_{r's'} \subset T_{NR}'$, where C_5 and C_5' are exchangeable, r and r' are end-vertices and $M_r \cap M_{r'} \neq \emptyset$.

Proof: Let $C_5 = \cup_{i=1}^2 E_{s_1 r_i} \cup E_{s_1 s_2} \cup_{i=3}^4 E_{s_2 r_i} \subset T_{NR}$, and $C_5' = E_{s_1' r_1'} \cup E_{s_1' r_4'} \cup E_{s_1' s_2'} \cup_{i=2}^3 E_{s_2' r_i} \subset T_{NR}$, where $M_{r_i} \cap M_{r_i'} \neq \emptyset$ $(i = 1, ..., 4)$. If $E_{s_1 r_1}$ or $E_{s_1' r_1'}$ is removed from C_5 or C_5', $C_3 = C_5|_{E_{s_1 r_1}^{max} \to E_{s_1 r_1}} - E_{s_1 r_1}^{max} = \cup_{i=2}^4 E_{s_2 r_i''}$ $(r_i'' \in M_{r_i} \cup M_{r_i'})$ or $C_3' = C_5'|_{E_{s_1' r_1'}^{max} \to E_{s_1' r_1'}} - E_{s_1' r_1'}^{max} = \cup_{i=2}^4 E_{s_2' r_i'''}$ $(r_i''' \in M_i \cup M_{r_i'})$.

According to Corollary 8.7.3, only one clique is enclosed by the minimum rectangle. Therefore, $C_3 = C_3' = C_5 - E_{s_1 r_1} = C_3' - E_{s_1' r_1'} = \cup_{i=2}^4 E_{sr_i''}$ (Figure 8.18).

When s is adjacent to r_1'', $s = \cap_{i=1}^4 E_{sr_i''}$. According to Corollary 8.3.3, $E_{sr_i''}$ $(i = 1, ..., 4)$ must be straight edge-sets. Therefore, the distance between r_1 or r_1' and $E_{sr_2''}$ equals the distance between r_1 or r_1' and $E_{sr_4''}$. According to Equation 8.8, $C_5' = (C_5 - E_{s_1 r_1}) \sqcup E_{s_1' r_1'}$. □

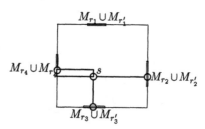

Figure 8.18. There is only one C_3 for three moving Steiner points

If $\{r_i | i = 1, ..., 4\} \subset V_s(T_{NR})$ and $\{r_i' | i = 1, ..., 4\} \subset V_s(T_{NR}')$, then $T_{NR} - C_5 = T_{NR}' - C_5'$ and $T_{NR} - E_{sr_i} = T_{NR}' - E_{s'r_i'}$, where $E_{sr_i} \in C_5$ and $E_{s'r_i'} \in C_5'$. Therefore, replacing a straight edge-set in C_5 is equivalent to replacing C_5.

Corollary 8.10.1. If $C_5 \in T_{NR}$ and $C_5' \in T_{NR}'$ are exchangeable, or $E_{rs} \in C_5$ and $E_{r's'} \in C_5'$ are exchangeable, then $T_{NR}' = (T_{NR} - C_5) \cup C_5' = (T_{NR} - E_{rs}) \cup E_{r's'}$, where r and r' are end-vertices and $M_r \cap M_{r'} \neq \emptyset$.

Example: Figure 8.19 demonstrates that C_5 in Figure 8.14 can be changed to its exchangeable clique by replacing a straight edge-set adjacent to an end-vertex with its exchangeable edge-set. Removing $E_{s_3 s_4}$ from T_{NR} in Figure 8.14 yields two NR edge-set subtrees (see Figure 8.19a). Because $E_{n_3 n_2} \natural E_{s_5 s_6} = \{M_{s_3}, M_{s_4} | x_{s_3} = x_{s_4}, |M_{s_3}| = |M_{s_4}| = 2\}$ (see Figure 8.19b), $s_3 \in P_{n_3 s_2} \cap P_{s_3 s_4}$

and $s_4 \in P_{s_5 s_6} \cap P_{s_3 s_4}$, s_4 $(x_{s_4} \geq x_{s_5})$ can be moved along the horizontal direction. Because $E_{s_3 s_4} \natural E_{n_4 n_5} = \{M_{s_4}, M_{s_6} \mid y_{s_4} = y_{s_6}, |M_{s_4}| = |M_{s_6}| > 1\}$, $E_{s_4 s_6}$ $(y_{s_4} = y_{s_6} \geq y_{s_5}$ can be moved along the vertical direction (Figure 8.19b). According to Theorem 8.4, $E_{s_5 s_7} \notin D_{s_4}$, because $D_{s_4} = \{E_{s_3 s_4}, E_{s_4 s_5}, E_{s_4 s_6}\}$. Therefore, $M_{s_4} = P_{fg}$, where $f = (x_{n_7}, y_{n_8})$ and $g = (x_{n_3}, y_{n_4})$. Similarly, $M_{s_5} = M_{s_4}$ can be obtained.

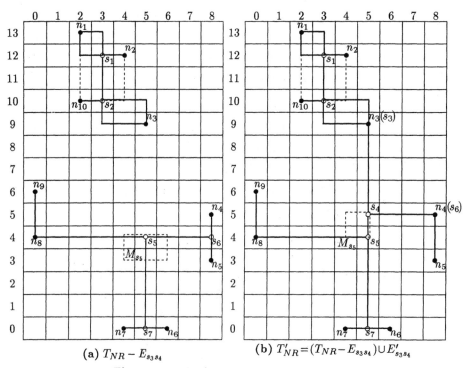

(a) $T_{NR} - E_{s_3 s_4}$ (b) $T'_{NR} = (T_{NR} - E_{s_3 s_4}) \cup E'_{s_3 s_4}$

Figure 8.19. Replacing an edge-set in T_{NR}

◇

According to Theorem 8.10 and Corollary 8.10.1, operations with "$-$" and "$\sqcup$" as operators obey the associativity that can be expressed as $T'_{NR} = (T_{NR} - C_5) \sqcup C'_5 = (T_{NR} - C_5) \sqcup (C_5 - E_{rs}) \sqcup E_{r's'} = \{[(T_{NR} - C_5) \sqcup C_5] - E_{rs}\} \sqcup E_{r's'} = (T_{NR} - E_{rs}) \sqcup E_{r's'}$. Furthermore, edge-set trees obtained by exchanging edge-sets in $C_5 = E_{s_1 s_2} \cup D_{E_{s_1 s_2}}$ and $C'_5 = E_{s'_1 s'_2} \cup D_{E_{s'_1 s'_2}}$ with their exchangeable edge-sets, not in C_5 and C'_5, can be obtained by exchanging edge-sets in $E_{s_1 s_2} \cup C'_5$ or $E_{s'_1 s'_2} \cup C_5$.

According to Corollary 8.1.1, $E_{pq} \cup D_{E_{pq}}$ can be exchanged with $E_{p'q'} \cup D_{E_{p'q'}}$, if $l_{E_{pq}} = l_{E_{p'q'}}$ for parallel moving edge-sets E_{pq} and $E_{p'q'}$. To reflect the exchangeable relation between $E_{pq} \cup D_{E_{pq}}$ and $E_{p'q'} \cup D_{E_{p'q'}}$, a parallel moving edge-set and its dependent edge-sets are considered a compound clique, denoted as C_{5p}. Note that $l_{C_{5p}}$ is constant, only if four end-vertices are fixed. Because $D_{E_{pq}} \neq D_{E_{p'q'}}$, $C_p - E_{pq} \neq C'_p - E_{p'q'}$. If the two-vertex moving edge-set in C_5 is replaced with a parallel moving edge-set, then two straight edge-sets in C_5 can be changed to bend edge-sets. Therefore, C_{5p} is also a special case of C_5. Corollary 8.1.1 has described the condition for the existence of an exchangeable clique for C_{5p}. Theorem 8.11 and Corollary 8.11.1 describe conditions for exchanging two compound cliques $C_{p_1} \subset T_{NR}$ and $C_{p_2} \subset T_{NR}$ with their exchangeable cliques.

Theorem 8.11. If $|P(C_{5p_1}) \cap P(C_{5p_2})| > 1$ for $C_{5p_1} \cup C_{5p_2} \subset T_{NR}$, and C'_{5p_1} and C'_{5p_2} are exchangeable cliques of C_{5p_1} and C_{5p_2}, respectively, then $l_{T_{NR}} < l_T$, where $T = [(T_{NR} - C_{5p_1}) \sqcup C'_{5p_1} - C_{5p_2}] \sqcup C'_{5p_2}$.
Proof: Because $|P(C_{5p_1}) \cap P(C_{5p_2})| > 1$, $l_{(C_{5p_1} \cup C_{5p_2})} = l_{C_{5p_1}} + l_{C_{5p_2}} - [l_{(C_{5p_1} \cap C_{5p_2})} + 1]$. In $T = [(T_{NR} - C_{5p_1}) \sqcup C'_{5p_1} - C_{5p_2}] \sqcup C'_{5p_2}$, $l_{(C'_{5p_1} \cup C'_{5p_2})} = l_{C'_{5p_1}} + l_{C'_{5p_2}} = l_{C_{5p_1}} + l_{C_{5p_2}} > l_{(C_{5p_1} \cup C_{5p_2})} + 1$. Therefore, $l_T = l_{T_{NR}} - l_{(C_{5p_1} \cup C_{5p_2})} + l_{(C'_{5p_1} \cup C'_{5p_2})} > l_{T_{NR}}$. $\square$

Example: In Figure 8.20a, $C_{5p_1} = E_{pq} \cup D_{E_{pq}}$ and $C_{5p_2} = E_{rs} \cup D_{E_{rs}}$ have their exchangeable compound cliques. Because $C_{5p_1} \cap C_{5p_2} = E_{qs}$, exchanging C_{5p_1} with its exchangeable clique reduces points in M_s and M_r, and the exchangeable clique for C_{5p_2} in Figure 8.20a no longer exists for C_{5p_2}, as shown in Figure 8.20b.

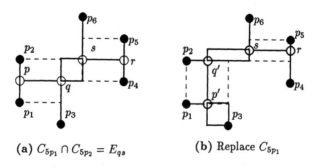

(a) $C_{5p_1} \cap C_{5p_2} = E_{qs}$ (b) Replace C_{5p_1}

Figure 8.20. C_{5p_1} and C_{5p_2} for $|P(C_{5p_1}) \cap P(C_{5p_2})| > 1$

Corollary 8.11.1. If $|P(C_{5p_1}) \cap P(C_{5p_2})| = 0$ or 1, then $l_{T'_{NR}} = l_{T_{NR}}$ for $T'_{NR} = [(T_{NR} - C_{5p_1}) \sqcup C'_{5p_1} - C_{5p_2}] \sqcup C'_{5p_2}$, where $C_{5p_1} \cup C_{5p_2} \subset T_{NR}$, and C'_{5p_1} and C'_{5p_2} are exchangeable cliques of C_{5p_1} and C_{5p_2}, respectively.

Theorem 8.12 describes the relationship between exchangeable compound cliques and exchangeable edge-sets in the compound cliques. Corollary 8.12.1 describes the duplicated edge-set trees obtained by exchanging compound cliques, or edge-sets of the compound cliques.

Theorem 8.12. $E_{ap} \in C_{5p}$ and $E_{ap'} \in C'_{5p}$ are exchangeable, if and only if C_{5p} and C'_{5p} are exchangeable, where a is an end-vertex of C_{5p} and C'_{5p}.
Proof: Let $C_{5p} = E_{ap} \cup E_{a'p} \cup E_{pq} \cup E_{bq} \cup E_{b'q}$ and $C'_{5p} = E_{a'q'} \cup E_{bq'} \cup E_{p'q'} \cup E_{ap'} \cup E_{b'p'}$ be exchangeable (see Figure 8.21). According to Corollary 8.1.1, C_{5p} and C'_{5p} are exchangeable, if and only if $l_{E_{pq}} = l_{E_{p'q'}}$, or $R_{a'b'}$ is a square. Furthermore, $l_{E_{ap}^{max}} = l_{E_{ap'}^{max}} = l_{E_{aa'}} = l_{E_{ab'}}$, if and only if $l_{E_{pq}} = l_{E_{p'q'}}$, or $R_{a'b'}$ is a square. Therefore, E_{ap} and $E_{ap'}$ are exchangeable, if and only if C_{5p} and C'_{5p} are exchangeable. $\square$

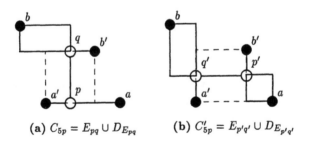

(a) $C_{5p} = E_{pq} \cup D_{E_{pq}}$ (b) $C'_{5p} = E_{p'q'} \cup D_{E_{p'q'}}$

Figure 8.21. C_{5p} is a special case of C_5

Corollary 8.12.1. If $C_{5p} \in T_{NR}$ and $C'_{5p} \in T'_{NR}$ are exchangeable, or $E_{ap} \in C_{5p}$ and $E_{ap'} \in C'_{5p}$ are exchangeable, then $T'_{NR} = (T_{NR} - C_{5p}) \sqcup C'_{5p} = (T_{NR} - E_{ap}) \sqcup E_{ap'}$, where a is an end-vertex of C_{5p} and C'_{5p}.

According to Theorem 8.12 and Corollary 8.12.1, all edge-set trees obtained by exchanging edge-sets in $C_{5p} = E_{pq} \cup D_{E_{pq}}$ and $C'_{5p} = E_{p'q'} \cup D_{E_{p'q'}}$ with their exchangeable edge-sets not in C_{5p} and C'_{5p} can be obtained by exchanging edge-sets in $E_{p'q'} \cup C_{5p}$ or $E_{pq} \cup C'_{5p}$ with their exchangeable edge-sets.

8.5   General Solution

If T'_{NR} is an arbitrary NR edge-set tree, generated by replacing edge-sets and cliques in T_{NR}, or in its exchangeable edge-set tree, with their exchangeable edge-sets and cliques, then T'_{NR} is called an equivalent edge-set tree of T_{NR}. The relationship between the general solution and the equivalent no-redundancy edge-set trees is described in Theorem 8.13.

Theorem 8.13. The general solution of Steiner's problem is contained by all equivalent edge-set trees for a set of nodes, where each NR edge-set tree is equivalent to all other NR edge-set trees. (Proof omitted.)

If $E_{p_i q_i} \in T_{NR} - E_{p_j q_j}$ $(i \neq j)$ for $T_{NR} = \cup_{i=1}^{n_T} E_{p_i q_i}$, then the total number of exchangeable edge-set trees, n_{exch_tree}, obtained from exchanging $E_{p_i q_i}$ and $E_{p_j q_j}$ with their exchangeable edge-sets, can be computed by:

$$n_{exch\_tree} = (n_i + 1)(n_j + 1) \quad (i \neq j) \tag{8.23}$$

where n_i is the number of exchangeable edge-sets of $E_{p_i q_i}$, and n_j is the number of exchangeable edge-sets of $E_{p_j q_j}$. If $E_{p_i q_i} \notin T_{NR} - E_{p_j q_j}$ $(i \neq j)$, then

$$n_{exch\_tree} = 1 + n_i + n_j \quad (i \neq j) \tag{8.24}$$

Based on the Theorems, Corollaries, and Equations, the general solution for Steiner's problem can be divided into following steps:

1. Find an original T_{NR} for V_n, using Equations 8.20 to 8.22.

2. Identify cliques from the dimension of every Steiner point in T_{NR}, using Theorem 8.1 and Corollaries 8.6.1 to 8.6.5 and 8.8.2 to 8.8.4.

3. Determine an exchangeable clique for every clique in T_{NR}, using Corollary 8.1.1, Theorem 8.7, Corollaries of Theorem 8.7, Theorem 8.11 and Theorem 8.12.

4. Determine exchangeable edge-sets and find all edge-set trees by replacing edge-sets in T_{NR} with their exchangeable edge-sets, using Equations 8.20 to 8.22.

Finding other exchangeable edge-set trees involves in more complicative computations and is omitted in this chapter. If redundant length is found in an edge-set tree, a new edge-set tree with shorter tree length can be obtained by eliminating the redundant length, and is used to replace the original T_{NR}. Then, above steps are repeated until all NR edge-set trees are equivalent.

If every moving Steiner point is given a specific location, all edge-sets in T_{NR} can be considered to be fixed edge-sets. Because there is only one edge in a fixed straight edge-set, the total number of MRSTs contained in T_{NR} is determined not only by each bend edge-set, but also by the domain of every dynamic Steiner point in T_{NR}. An edge-set tree that is obtained by shrinking every straight edge-set of T_{NR} into a point is called a bend edge-set tree, and is denoted as T_B.

Example: The bend edge-set tree in Figure 8.22 is obtained by shrinking every straight edge-set of T_{NR} in Figure 3.14 into a point.

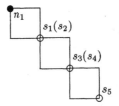

Figure 8.22. A bend edge-set tree

◇

Let $\{E_{p_1,q_1}, ..., E_{p_{n_b},q_{n_b}}\}$ be a set of all bend edge-sets, and $\{s_1, ..., s_{n_s}\}$ be a set of moving Steiner points in T_B, where $M_{s_i} = P_{f_i g_i}$ ($p_i = (x_i, y_i)$, $g_i = (x_i', y_i')$, $x_i \le x_i'$ and $y_i \le y_i'$). The total number of MRSTs contained by an NR edge-set tree, N_{MRST}, is computed by:

$$N_{MRST} = \prod_{i=1}^{n_s} \sum_{x_{s_i}=x_i}^{x_i'} \sum_{y_{s_i}=y_i}^{y_i'} \prod_{j=1}^{n_b} A_0(p_j, q_j) \qquad (8.25)$$

where $\prod_{i=1}^{n_s} \sum_{x_{s_i}=x_i}^{x_i'} \sum_{y_{s_i}=y_i}^{y_i'} = \sum_{x_{s_1}=x_1}^{x_1'} \sum_{y_{s_1}=y_1}^{y_1'} ... \sum_{x_{s_{n_s}}=x_{n_s}}^{x_{n_s}'} \sum_{y_{s_{n_s}}=y_{n_s}}^{y_{n_s}'}$, and $A_0(p_j, q_j)$ is computed by Equation 8.3. If all edge-sets in T_{NR} are straight edge-sets, then

$$N_{MRST} = \prod_{i=1}^{n_s}(x_i' - x_i + 1)(y_i' - y_i + 1) \qquad (8.26)$$

Example: For the NR edge-set tree in Figure 8.4a, $T_B = E_{n_1 n_2} \cup E_{n_2 s_1} \cup E_{s_2 n_5}$, where $2 \le x_{s_1} = x_{s_2} \le 3$, $n_1 = (0,0)$, $n_2 = (1,1)$, $s_1 = (x_{s_1}, 2)$, $s_2 = (x_{s_2}, 4)$, and $n_5 = (2,5)$. $N_{MRST} = \sum_{x=2}^{4} A_0(n_1, n_2) A_0(n_2, s_1) A_0(s_2, n_5) = A((0,0),(1,1)) \sum_{x=2}^{4} A_0((1,1),(x,2)) A_0((x,4),(2,5)) = C(2,1)[C(2,1)C(1,1) + C(3,1)C(2,1) + C(4,1)C(3,1)] = 2(2+6+2) = 40$, where $x = x_{s_1} = x_{s_2}$.

For T_{NR} in Figure 8.14, $T_B = E_{n_1 s_1} \cup E_{s_2 s_3} \cup E_{s_4 s_5}$, where $2 \leq x_{s_1} = x_{s_2} \leq x_{s_3} = x_{s_4} \leq 5$, $x_{s_4} \leq x_{s_5} \leq 6$, $y_{s_1} = 12$, $y_{s_2} = 10$, $y_{s_3} = 9$, $4 \leq y_{s_4} \leq 6$ and $3 \leq y_{s_5} \leq 5$. $N_{MRST} = \sum_{x_{s_1}=2}^{4} \sum_{x_{s_4}=x_{s_1}}^{5} \sum_{y_{s_5}=3}^{5} \sum_{x_{s_5}=max(4,x_{s_4})}^{6} \sum_{y_{s_4}=max(4,y_{s_5})}^{6}$ $A_0'(2, 13, x_{s_1}, 12) \, A_0'(x_{s_1}, 10, x_{s_4}, 9) \, A_0'(x_{s_4}, y_{s_4}, x_{s_5}, y_{s_5}) = 1595$.

$\diamond$

8.6 Procedures

The distance between E_{fg} and $E_{f'g'}$ is the distance between the two nearest edges of E_{fg} and $E_{f'g'}$ if $E_{fg} \natural E_{f'g'} \neq \emptyset$, or the distance between the two nearest vertices of E_{fg} and $E_{f'g'}$ if $E_{fg} \natural E_{f'g'} = \emptyset$. Expressing the distance between E_{fg}^{max} and $E_{f'g'}^{max}$ as $d(E_{fg}, E_{f'g'})$, $E_{ss'}^{min}$ with the length $l_{E_{ss'}^{min}} = d(E_{fg}, E_{f'g'})$ can be determined, where $P(E_{ss'}^{min}) \cap P(E_{fg}^{max}) = s$ and $P(E_{ss'}^{min}) \cap P(E_{f'g'}^{max}) = s'$. If $E_{fg}^{max} \natural E_{f'g'}^{max} = \{M_s, M_{s'} \mid x_s = x_{s'} \text{ or } y_s = y_{s'}, |R_s| = |R_{s'}| \geq 1\}$, then $E_{ss'}^{min}$ is a straight edge-set or a parallel moving edge-set (Theorem 8.1). If $E_{fg}^{max} \natural E_{f'g'}^{max} = \emptyset$, then $E_{ss'}^{min}$ is a bend edge-set (Corollary 8.1.2). If s or s' is not a node, then s or s' is a Steiner point. Based on the computation for $E_{ss'}^{min}$, Equation 8.21 can be implemented.

According to Corollaries 8.6.3 and 8.6.6, $E_{ss'}$ can be obtained from $E_{ss'}^{min}$ by maximizing $|M_s|$ and $|M_{s'}|$. Let $D_s = \{E_{ss'}, E_{sf}, E_{sg}\}$, $D_f = \{E_{sf}, E_{ff'}, E_{ff''}\}$, $D_g = \{E_{sg}, E_{gg'}, E_{gg''}\}$, and $s \notin V_n$, $|M_s|$ can be maximized by calling:

$FUNCTION\ M_s(E_{ss'}, D_f, D_g);$
$BEGIN$
    $IF\ E_{ss'}\ is\ a\ bend\ edge\text{-}set,\ THEN\ BEGIN$
        $IF\ E_{ss'}\natural E_{ff'} = \{s, f\}\ THEN\ \{e_{p_1 p_2}, e_{f_1 f_2}\} = E_{ss'}\natural E_{ff''}$
        $ELSE\ \{e_{p_1 p_2}, e_{f_1 f_2}\} = E_{ss'}\natural E_{ff'};$
        $IF\ E_{ss'}\natural E_{gg'} = \{s, g\}\ THEN\ \{e_{p_1' p_2'}, e_{g_1 g_2}\} = E_{ss'}\natural E_{gg''}$
        $ELSE\ \{e_{p_1' p_2'}, e_{g_1 g_2}\} = E_{ss'}\natural E_{gg'}$
    $END\ ELSE$
        $IF\ E_{pf}\ is\ a\ straight\ edge\text{-}set\ AND\ M_s = P_{p_1' p_2'} \subset P_{pf}\ THEN\ BEGIN$
        $IF\ E_{ss'}\natural E_{ff'} = \{s, f\}\ THEN\ \{e_{p_1 p_2}, e_{f_1 f_2}\} = E_{ss'}\natural E_{ff}'$
        $ELSE\ \{e_{p_1 p_2}\}, e_{f_1 f_2}\} = E_{ss'}\natural E_{ff}'$
    $END;$
    $x_a = min(x_{p_1}, x_{p_2}, x_{p_1'}, x_{p_2'});$
    $y_a = min(y_{p_1}, y_{p_2}, y_{p_1'}, y_{p_2'});$

$$x_b = max(x_{p_1}, x_{p_2}, x_{p'_1}, x_{p'_2});$$
$$y_b = max(y_{p_1}, y_{p_2}, y_{p'_1}, y_{p'_2});$$
$$M_s = P_{ab}$$
END.

Based on *FUNCTION* M_s, Equation 8.22 can be implemented, and T_{NR} can be obtained from an arbitrary edge-set tree, $T = \cup_{i=1}^{n_T} E_{p_i q_i}$, by calling:

FUNCTION $T_{NR}(T)$;
BEGIN
> $T_{NR} = T$; *i=1*;
> *REPEAT*
>> $\{T_1, T_2\} = T_{NR} - E_{p_i q_i}$;
>> *IF* $P_{f_g} \cap P(T_1) = f$ *AND* $P_{f_g} \cap P(T_2) = g$ *AND* $l_{E_{f_g}} < l_{E_{p_i q_i}}$ *THEN BEGIN*
>>> $T_{NR} = T_1 \sqcup E_{f_g} \sqcup T_2$;
>>> $n_T = $ *the number of edge-sets in* T_{NR};
>>> $i = 0$;
>> *END*
>> *ELSE IF* $M_f \supset M_{p_i}$ *AND* $M_g \supset M_{q_i}$ *are found THEN BEGIN*
>>> $M_{p_i} = M_f$;
>>> $M_{q_i} = M_g$
>> *END*;
>> *INC(i)*
> *UNTIL* $i > n_T$
END.

If the numbers of edge-sets in T_1 and T_2 are n_{T_1} and n_{T_2}, respectively, then the time needed to find E_{f_g} is proportional to $n_{T_1} \cdot n_{T_2}$. According to Corollary 8.3.6, the average number of Steiner points is $(n-2)/2$, where $n = |V_n|$, and the average number of edge-sets is $n + (n-2)/2 - 1 = 3n/2 - 2$. Considering an edge-set is deleted, the time to find E_{f_g} is approximately proportional to $\sum_{i=1}^{3n/2-4} i(3n/2 - 3 - i)/(3n/2 - 2) = (n-2)(3n-8)/8$ ($n \geq 3$), and the time needed to find T_{NR} is approximately proportional to $\frac{1}{8}(n-2)(3n-8)(3n/2-2) = \frac{1}{16}(n-2)(3n-4)(3n-8)$ ($n \geq 3$).

If no redundant edge-set exists in adjacent edge-sets in the initial edge-set tree T, then the time needed to find T_{NR} by *FUNCTION* T_{NR} is reduced. The following function can be used to find T:

FUNCTION $T(V_n)$;
BEGIN
    $T = E_{n_1 n_2}; i = 2;$
    WHILE $i \neq |V_n|$ DO BEGIN
        INC(i);
        Find $E_{pq} \in T$ so that $d(E_{pq}, E_{n_i n_i})$ is the minimum;
        $T = T \cup (E_{pq} + n_i);$
    END
END.

In *FUNCTION T*, the arithmetic of $E_{pq} + n_i$ is defined in Equation 8.8.

Finding exchangeable edge-sets for every edge-set in T_{NR} is similar to finding an edge-set that connects two separated edge-set trees with the minimum length. Let $X_i(E_{p_i q_i})$ be a stack containing exchangeable edge-sets of $E_{p_i q_i} \in T_{NR}$, then $X_i(E_{p_i q_i})$ can be found by the following function.

FUNCTION $X_i(E_{p_i q_i})$;
BEGIN
    $\{T_1, T_2\} = T_{NR} - E_{p_i q_i};$
    IF $l_{E_{f_j g_j}} = l_{E_{p_i q_i}}$ $(j = 1, ..., n)$
    in which $E_{f_j g_j} \neq E_{p_i q_i}$, $f = E_{p_j q_j} \cap T_1$ and $g = E_{f_j g_j} \cap T_2$, THEN
        $X_i = \{E_{f_1 g_1}, ..., E_{f_n g_n}\}$
END.

8.7 Example

In this example, the node configuration, $V_n = \{n_1, n_2, ..., n_{10}\}$, shown in Figure 8.14 is selected to explain how to generate T_{NR} and its equivalent trees which contain the general solution. Initially, *FUNCTION $T(V_n)$* is called for generating an edge-set tree with all Steiner points fixed. First, nodes n_1 and n_2 are connected and stored as the edge-set tree T: $T = E_{n_1 n_2}$. Then, node n_3 is connected to T: $T = T \cup E_{n_2 n_3}$. Because no Steiner point is generated, in this stage, $T = E_{n_1 n_2} \cup E_{n_2 n_3}$. Similarly, nodes n_4 and n_5 are connected to T: $T = T \cup E_{n_3 n_4} = E_{n_1 n_2} \cup E_{n_2 n_3} \cup E_{n_3 n_4}$, and $T = T \cup E_{n_4 n_5} = E_{n_1 n_2} \cup E_{n_2 n_3} \cup E_{n_3 n_4} \cup E_{n_4 n_5}$. Because n_6

has the same distance to $E_{n_3n_4}$ and $E_{n_4n_5}$, n_6 can be connected to $E_{n_3n_4}$ or $E_{n_4n_5}$. Here, n_6 is connected to $E_{n_3n_4}$. Because Steiner point s_5 is needed to connect n_3, n_4 and n_6 with the minimum length, the arithmetic expressed by Equation 8.8 is used to split $E_{n_3n_4}$ into two: $T = T \sqcup E_{n_6s_5} = (T - E_{n_3n_4}) \cup E_{n_3s_5} \cup E_{n_4s_5} \cup E_{n_6s_5} = E_{n_1n_2} \cup E_{n_2n_3} \cup E_{n_3s_5} \cup E_{n_4s_5} \cup E_{n_4n_5}$. Repeating the process as above, nodes n_7, ..., n_{10} can be connected as follows: $T = T \sqcup E_{n_6n_7}$, $T = T \sqcup E_{n_8s_4} = (T - E_{n_3s_5}) \cup E_{n_3s_4} \cup E_{s_4s_5} \cup E_{n_8s_4}$, $T = T \sqcup E_{n_9s_8} = (T - E_{n_8s_4}) \cup E_{n_8s_8} \cup E_{s_4s_8} \cup E_{n_9s_8}$, and finally $T = T \sqcup E_{n_{10}s_1} = (T - E_{n_1n_2}) \cup E_{n_1s_1} \cup E_{n_2s_1} \cup E_{n_{10}s_1}$ which is shown in Figure 8.23a.

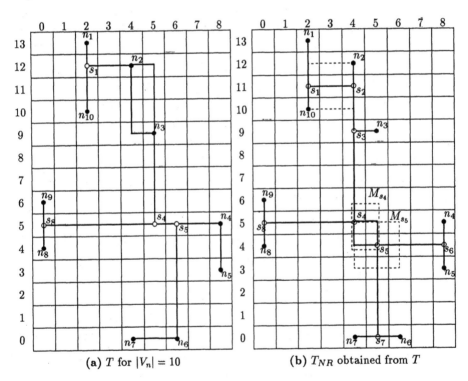

(a) T for $|V_n| = 10$ (b) T_{NR} obtained from T

Figure 8.23. An initial edge-set tree and an original NR edge-set tree

In order to change the edge-set tree in Figure 8.23a to an original NR edge-set tree, domains of the Steiner points in T must be maximized by examining the overlap between two edge-set and calling *FUNCTION* M_s. Because a vertical overlap exists between $E_{n_{10}s_1}$ and $E_{n_2n_3}$, $E_{n_{10}s_1} \natural E_{n_2n_3} = \{M_{s_1} = e_{n_{10}s_1}, M_{s_2} = e_{n_2a_1}\}$, where $a_1 = (4,10)$, edge-set $E_{n_2 s_1}$ must replaced by the parallel moving edge-set $E_{s_1 s_2}$: $T = (T - E_{n_2 s_1}) \sqcup E_{s_1 s_2} = (T - E_{n_2 s_1} - E_{n_1 n_{10}} - E_{n_2 n_3}) \cup E_{n_1 s_1} \cup E_{n_{10} s_1} \cup E_{n_2 s_2} \cup E_{n_3 s_2} \cup E_{s_1 s_2}$. Similarly, an overlap between $E_{n_3 s_2}$ and $E_{s_4 s_8}$ is found: $E_{n_3 s_2} \natural E_{s_4 s_8} = \{M_{s_3} = e_{n_3 a_2}, M_{s_4} = e_{b_1 b_1'}\}$, where $a_2 = (4,9), b_1 = (4,5)$, and $b_1' = (5,5)$. Edge-set $E_{n_3 n_4}$ must be replaced by a parallel moving edge-set $E_{s_3 s_4}$: $T = (T - E_{n_3 s_4}) \sqcup E_{s_3 s_4} = [(T - E_{n_3 s_4}) \cup E_{s_3 s_4} - E_{n_3 s_2}] \cup E_{s_2 s_3} \cup E_{n_3 s_3} \cup E_{s_3 s_4}$. By the use of FUNCTION M_s, the domain of s_4 can be maximized: $M_{s_4} = M_{s_4}(E_{n_3 s_4}, E_{s_4 s_8}, D_{s_4 s_8}) = P_{b_2 b_1''}$, $M_{s_8} = P_{c_1 n_9}$, where $b_1'' = (5,6)$, and $c_1 = (0,5)$. In this stage, M_{s_4} is enclosed by a rectangle. Repetitively examining overlaps and using FUNCTION M_s, new Steiner points can be found, and the domain of each Steiner point can be maximized. Because $E_{n_4 n_5} \natural E_{n_6 s_5} = \{M_{s_5} = e_{a_3 a_3'}, M_{s_6} = e_{n_4 n_5}\}$, where $a_3 = (6,3)$, and $a_3' = (6,5)$, $E_{n_4 n_5}$ must be replaced by the parallel moving edge-set $E_{s_5 s_6}$: $T = (T - E_{n_4 s_5}) \cup E_{s_5 s_6} = ((T - E_{n_4 s_5}) \cup E_{s_5 s_6} - E_{n_4 n_5} - E_{n_6 n_7}) \cup E_{n_4 s_6} \cup E_{n_5 s_6} \cup E_{n_6 s_7} \cup E_{n_7 s_7}$, where $M_{s_5} = M_{s_5}(E_{n_4 s_5}, D_{s_4}, D_{n_6}) = P_{a_3'' a_3'}$ $(a_3'' = (4,3))$, $M_{s_6} = P_{n_4 n_5}$, and $M_{s_7} = P_{n_6 n_7}$. Because $E_{n_8 n_9} \natural E_{s_4 s_5} = \{M_{s_5} = e_{a_4 a_4'}, M_{s_8} = e_{n_8 n_9}\}$, where $a_4 = (4,4)$, and $a_4' = (4,6)$, $M_{s_4} = M_{s_4}(E_{s_4 s_5}, D_{s_3}, D_{s_8}) = P_{b_2''' b_2''}$, where $b_2''' = (4,4)$, and $M_{s_8} = P_{n_8 n_9}$. After the domain of each Steiner point is maximized, the edge-set tree is shown in Figure 8.23b. This edge-set tree is an original edge-set tree, which can be used to further generate exchangeable edge-set trees, because no redundant length can be found after examining by FUNCTION $T_{NR}(T)$.

In order to generate all exchangeable edge-set trees from an NR edge-set tree, exchangeable edge-sets/cliques of every edge-set/clique in this edge-set tree have to be determined. Because $|M_{s_4} \cap M_{s_5}| > 1$, $l_{E_{s_5 s_7}^{max}} = l_{E_{n_5 n_6}}$ and $l_{E_{s_1 s_2}} = |M_{s_1}|$ in T_{NR} (see Figure 8.23b), $C_5' = E_{s_4' s_5'} \cup D_{E_{s_4' s_5'}}$, $E_{n_5 n_6}$, and $C_{5p}' = E_{s_1' s_2'} \cup D_{E_{s_1' s_2'}}$ can be exchanged with $C_5 = E_{s_4 s_5} \cup D_{E_{s_4 s_5}}$, $E_{s_5 s_7}$ and $C_{5p} = E_{s_1 s_2} \cup D_{E_{s_1 s_2}}$, respectively. If $x_{s_5} = 5$, then $l_{E_{s_5 s_6}} = |M_{s_5}| = 3$. It seems there exists an exchangeable clique for $E_{s_5 s_6} \cup D_{E_{s_5 s_6}}$. According to Theorem 8.12, exchanging $E_{s_5 s_6} \cup D_{E_{s_5 s_6}}$ with $E_{s_5' s_6'} \cup D_{E_{s_5' s_6'}}$ cannot generate a new edge-set tree, because $E_{s_5 s_7} \in D_{E_{s_5 s_6}}$ and $E_{n_5 n_6} \in D_{E_{s_5' s_6'}}$ have previously been considered to be a pair of exchangeable edge-sets. Edge-set trees obtained by exchanging C_5, C_{5p_1}, $E_{s_5 s_7}$ with C_5', C_{5p_1}', $E_{n_5 n_6}$ are given as: $T_{NR}' = (T_{NR} - C_5) \cup C_5'$ (see Figure 8.24a), $T_{NR}'' = (T_{NR} - C_{5p_1}) \cup C_{5p_1}'$ (see Figure 8.14), and $T_{NR}''' = (T_{NR}' - C_{5p_1}) \cup C_{5p_1}' = (T_{NR}'' - C_5) \cup C_5'$ (see Figure 8.24b). Because $E_{s_5 s_7} \in T_{NR}$ and $E_{s_5 s_7} \in T_{NR}''$, edge-set trees obtained by exchanging edge-sets $E_{s_5 s_7}$ with $E_{n_5 n_6}$ are $T_{NR}^{(4)} = (T_{NR} - E_{s_5 s_7}) \cup E_{n_5 n_6}$, and

$T_{NR}^{(5)} = (T_{NR}'' - E_{s_5 s_7}) \cup E_{n_5 n_6}$ (see Figures 8.27c and 8.27d).

According to Corollaries 8.10.1 and 8.12.1, there may exist new exchangeable edge-sets for $E_{s_1' s_2'}$, or $E_{s_4' s_5'}$, or edge-sets in $D_{E_{s_1' s_2'}}$ or $D_{E_{s_4' s_5'}}$, that are not edge-sets in C_5, C_5', C_{5p} and C_{5p}'. However, there is no exchangeable edge-sets for $E_{s_1' s_2'}$. Because the domain of $s_4 \in T_{NR}''$ in Figure 8.14 is different from that of $s_4 \in T_{NR}$ in Figure 8.23b, $E_{s_4 s_5} \in T_{NR}''$ can be exchanged with $E_{n_3 n_4}$. Furthermore, $C_{p_2} = E_{s_3 s_4} \cup D_{E_{s_3 s_4}} \subset T_{NR}''$, and $E_{s_3 s_4} \in T_{NR}''$ can be exchanged with $C_{p_2}' = E_{s_3'' s_4''} \cup D_{E_{s_3'' s_4''}}$, and $E_{n_9 s_3}$, respectively. Exchanging $E_{s_4 s_5}$, C_{5p_2}, and $E_{s_3 s_4}$ with $E_{s_4' s_5'}$, C_{5p_2}', and $E_{n_9 s_3}$, respectively, yields exchangeable edge-set trees: $T_{NR}^{(6)} = (T_{NR}'' - E_{s_4 s_5}) \cup E_{n_3 n_4}$, $T_{NR}^{(7)} = (T_{NR}'' - C_{5p_2}) \cup C_{5p_2}'$, and $T_{NR}^{(8)} = (T_{NR}'' - E_{s_3 s_4}) \cup E_{n_9 s_3}$ (see Figures 8.27e to 8.27g). Edge-set trees T_{NR}, T_{NR}', ..., $T_{NR}^{(8)}$ can be proven to be equivalent by FUNCTION T_{NR}, because no redundancy exists in any one of them. Therefore, the general solution for V_n in Figure 8.23 is contained by the nine NR edge-set trees (see Figures 8.14, 8.23b and 8.24).

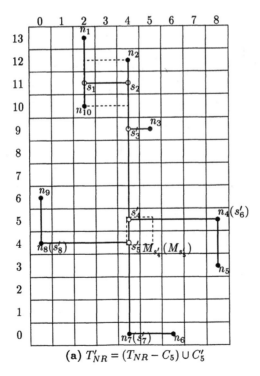

(a) $T_{NR}' = (T_{NR} - C_5) \cup C_5'$

Figure 8.24. Equivalent edge-set trees of T_{NR} (continue)

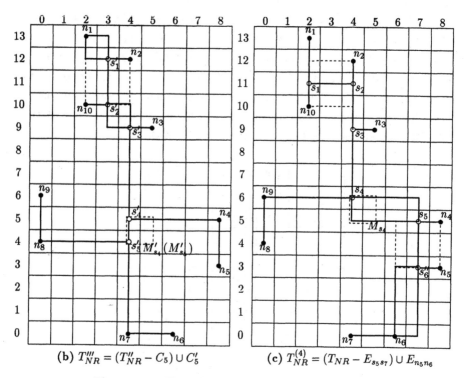

Figure 8.24. Equivalent edge-set trees of T_{NR} (continue)

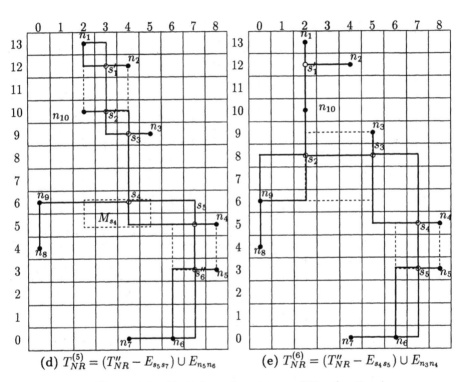

Figure 8.24. Equivalent edge-set trees of T_{NR} (continue)

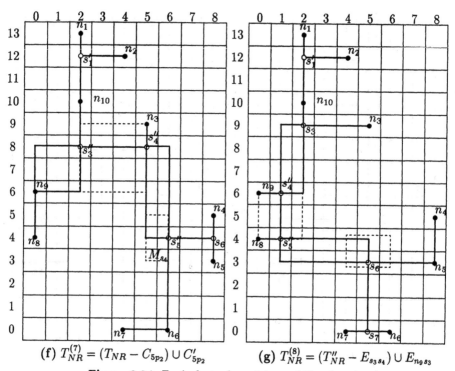

(f) $T_{NR}^{(7)} = (T_{NR} - C_{5p_2}) \cup C'_{5p_2}$ **(g)** $T_{NR}^{(8)} = (T''_{NR} - E_{s_3 s_4}) \cup E_{n_9 s_3}$

Figure 8.24. Equivalent edge-set trees of T_{NR} (continue)

Appendix A: Symbols

A_d: Total device area

A_{di}: Area of device i

A_p: Accessing planes

A_r: Routable substrate area

A_s: Aspect ratio of the substrate

A_{uf}: Area utilization factor

$A_0(p,q)$: Total number of alternative edges in E_{pq} when $|\Re_p| = |\Re_q| = 1$

C_{ai}: Accessing connectivity of device i

C_{di}: Demanded connectivity of device i

C_n: Clique with n edge-sets

C_{nd}: Degenerated clique with n edge-sets

C_{pi}: Permitted connectivity of device i

$C_{pi-outer}$: Permitted connectivity on the outer layer

$C_{pi-olayer}$: Permitted connectivity on the inner layers

C_{ti}: Interconnect connectivity of device i

C_y: Edge-set cycle consisting of edge-sets, that contains a set of cycles

C_{5p}: Clique with a parallel moving edge-set and its dependent edge-sets

D: Via-pad diameter

D_b: Board density

$deg\ s$: Degree of $s \in V_t$, the number of edge-sets that intersects at s

$D_{E_{pq}}$: Set of edge-sets that is always adjacent to E_{pq}

$D(E_{pq}, E_{fg})$: Distance between edge-sets E_{pq} and E_{fg}

D_p: Set of edge-sets that always intersect at p

D_{pi}: Lead-pad width/diameter

D_{pq}: Distance between $p = (x_p, y_p)$ and $q = (x_q, y_q)$, equal to $|x_p - x_q| + |y_p - y_q|$
for rectilinear distance, or $[(x_p - x_q)^2 + (y_p - y_q)^2]^{1/2}$ for Euclidean distance

D_s: Substrate density

D_{sv}: Signal via-pad diameter

D_{thi}: Diameter of the ith tooling hole

$E(G)$: Set of all edges in graph G

e_{pq}: Shortest rectilinear edge from p to q, $e_{pq} \in E_{pq}$

E_{pq}: Edge-set that contains a set of all shortest edges between points p and q

E_{pq}^{max} (or E_{pq}^{min}): Fixed edge-set in E_{pq} when the distance between p and q is the
maximum (or minimum)

E_T: Set of all edge-sets in an edge-set tree T

$E(v)$: Set of edges that intersect at v, which is used in graph partitioning

E_X: Error index for tree type X

F_{ij}: Force acting on module i by module j, used in the force-directed placement

$G = (E, V)$: Graph with edge set, $E = E(G)$, and vertice set, $V = V(G)$

I_{pr}: Interconnection to pin ratio

K_b: Bogatin's constant

K_c: Coomb's constant

K_m: Messner's constant

K_w: Welterlen's constant

L_{ave}: Average wire length per interconnection, or average number of signal layers

L_{cti}: Length of cutout i

L_{di}: Length of device i

$\bar{l}_{E_{pq}}$: Average length of edges contained in E_{pq}

$l_{R_{pq}}$: Perimeter of R_{pq}, $l_{R_{pq}} = 2l_{E_{pq}}$

l_x: Length of $x \in \{e_{pq}, E_{pq}, C_n, C_{5p}, C_d, C_y, t, T, T_{NR}\}$

$max(a, ..., z)$: Function that selects the maximum one in $\{a, ..., z\}$

$mid(a, b, c)$: Function that selects one with the mid-value in $\{a, b, c\}$

$min(a, ..., z)$: Function that selects the minimum one in $\{a, ..., z\}$

M_p: Domain of p, a set of points enclosed by a rectangle

N_{ct}: Number of cutouts

N_l: Total number of device leads

N_{li}: Number of leads in device i

n_t: Number of tracks per channel permitted in the substrate

n_{tai}: Number of tracks per channel required for accessing by device i

n_{ti}: Number of tracks per channel permitted by device i

N_{th}: Number of tooling holes

N_{tsv}: Number of traces per channel permitted by signal vias

N_w: Total number of signal interconnections

N_{wi}: Number of signal interconnections by device i

P: Pitch between two vias

P_{di}: Lead pitch of device i, or pitch between adjacent leads of device

P_{pq}: Set of points enclosed by a rectangle with p and q as vertices

P_{sv}: Pitch between two signal vias

P_T: Intersection of point-sets of all edge-sets in edge-set tree T

R_e: Routing efficiency

R_i: Number of rows from the periphery in device i

R_{pq}: Rectangle with p and q as vertices

S: Trace spacing

S_d: Pitch between two devices

S_{fi}: Shape factor of device i

T: Trace width

T_B: Bend-edge-set tree

T_{NR}: No-redundancy edge-set tree

T_{se}: Trace to substrate edge spacing

V_n: Set of given nodes

$V_s(T)$: Set of Steiner points in edge-set tree T for connecting nodes

V_T: Set of nodes and Steiner points, $V_t = V_T = V_n \cup V_s(T)$ if $t \in T$

W_{cti}: Width of cutout i

W_d: Demanded wire length

W_{di}: Width of device i

W_p: Permitted wire length

(x_r, y_r): Point r with x_r and y_r as horizontal and vertical coordinates

$|X|$: Number of elements in set X

$\natural$: Overlap operator expressing the related position of two separate edge-sets

$\sqcup$: Operator that functions as a union associated with breaking an edge-set into two edge-sets by a Steiner point

Appendix B: Acronyms

AUF: Area utilization factor

AXL: Axial resistor/capacitor

CNF: Conjunctive normal form

CVM: Constrained via minimization problem

DIP: Dual in-line package

ECONN: Edge card connector

EIC: Equivalent integrated circuit count

FEM: Finite element model

HCONN: High-density connector

IDEG: Iso-distance error graph

IDG: Iso-distance graph

MCM: Multichip Module

MLS: Module-lead-substrate

MOCC: Minimum odd cycle cover

MRST: Minimum rectilinear Steiner tree

MSPT: Minimum rectilinear Spanning tree

n-CVM: N-layer constrained via minimization problem

OCC: Odd cycle cover

PGA: Pin-grid array

PLCC: Plastic leaded chip carrier

PNC: Planar node cover problem

PWB: Printed wiring board

QFP: Quad-flat pack

TAB: Tape automated bond

TCM: Thermal conduction module

UVM: Unconstrained via minimization problem

VDB: Vertex-deletion graph bipartization problem

VDB4: VDB for planar graph with maximum degree of vertex limited to 4

2-CVM: Two-layer constrained via minimization problem

2-CVM4: 2-CVM for the maximum junction degree limited to 4

Appendix C: Glossary

Assignment: The grouping of modules to a workspace prior to placement.

Accessing connectivity: The length of wiring per unit area required for accessing the device pins from the device periphery.

Adjacent edges (edge-sets): Two edges (edge-sets) that intersect at a point (vertex).

Adjacent points (vertices): Two points (vertices) that are directly connected by an edge (edge-set).

Average connection error: The average value of connection errors generated by moving a node from one trace to another trace on an IDEG.

Backward search: A search process searching a path connecting two given terminals from the progressing information provided by the forward search.

Bend: An angle in an edge, especially a right angle in a rectilinear edge.

Bend edge: An edge with one or more bends.

Bend edge-set: A set of bend edges connecting two vertices.

Bend edge-set tree: A special edge-set tree generated by contracting all straight edge-sets in an edge-set tree to points, which is used to compute the number of trees contained by the edge-set tree.

Bipartite graph: A graph in which vertices are partitioned into two subsets, and each vertex of one subset is adjacent to every vertex of another subset.

Bounded length: The length difference of two paths bounded by the requirement for routing in high-speed circuits.

Chain: A tree in which no vertex (node) connects more then two edges.

Channel: A path between two via-pads.

Characteristic IDEG: The IDEG with the maximum error index for a set of nets that is to route in a given placement workspace.

Clearance: The insulation area between two neighboring parallel wires in a printed wiring circuit/board.

Clique: An edge-set subtree consisting of a set of moving edge-sets with the minimum total length independent of locations of dynamic Steiner points, and with the minimum number of edge-sets for maximizing the number of points in each domain.

Clustering-development placement: A constructive placement based on the average number of edges connecting placed and unplaced modules.

Combination: Taking r elements at a time from n $(n \geq r)$ distinct elements, without considering the order.

Complete graph: A graph in which each node is connected to all other nodes.

Connected graph: A graph in which each vertex is connected to any other vertices with one or more edges.

Connection error: The length difference generated by connecting a node to a tree with the length of both a non-MRST and an MRST, used for approximating MRSTs in placement.

Connection order: The order for connecting terminals one after another to form a net.

Connectivity: The number of nets connecting two modules in two subsets.

Constructive placement: A class of placement methods, in which unplaced modules are placed in terms of their connections to the placed modules, or all modules are simultaneously placed in locations determined by computation.

Correction function: A function associated with the weighted routing length for improving the placement configuration.

Cost: A value used to estimate the placement quality, as measured by one or more specific placement objectives.

Cost function: A function for computing the cost that determines the selection of a placement configuration.

Cost matrix: A matrix where every element is the cost between two modules.

Cycle: A sub-graph in which every vertex connects exactly two edges.

Degenerated clique: An interface edge-set subtree that connects non-degenerated cliques with given nodes, in which a dependent edge-set of each Steiner point is degenerated.

Degree of a vertex (node, or point): The number of edges (edge-sets) connected to a vertex in a graph (an edge-set tree).

Demanded connectivity: The signal wire length required per unit area of substrate.

Density: The routable substrate area available per device.

Dependent edge-set of an edge-set: Edge-set that is always adjacent to an edge-set in a no-redundancy edge-set tree.

Dependent edge-set of a point: Edge-set that always intersect at a point (node or Steiner point) in a no-redundancy edge-set tree.

Detailed routing: The final determination of interconnections on a substrate by searching a path connecting a set of terminals (pins) in a workspace, where the substrate associated with the routed paths is simulated.

Detour: An additional sub-path required for routing around obstacles to connect two terminals in detailed routing.

Distance matrix: A matrix where each element is the distance between two vertices in a graph.

Domain of a Steiner point: A set of points that defines the moving range of a Steiner point.

Dynamic Steiner point: A Steiner point that can be moved in a rectangular domain.

Edge: A connection of two vertices with specific measures of length and shape.

Edge-set: A set of edges that connects two vertices (points) with the minimum rectilinear distance.

Edge-set cycle: A graph in which each vertex connecting two edge-sets.

Edge-set tree: A graph generated by expanding every edge in a tree into an edge-set.

End-vertex of a clique: A vertex that is a node or a common Steiner point of two edge-sets in different cliques.

Equivalent edge-set tree: An no-redundancy edge-set tree that must not be converted to an edge-set tree containing redundancy, by replacing edge-sets with their exchangeable edge-sets.

Equivalent integrated count: A single 14-pin DIP.

Error index: The average connection error of an $IDEG$ produced by moving a node in an $IDEG$ from any trace to its adjacent lower trace, that is used to estimate how well a tree type approximates the minimum rectilinear Steiner tree in placement.

Euclidean distance: The distance of two vertices, that is measured by the length of a straight line-segment with the vertices as terminals.

Exchangeable edge-set: An edge-set that can replace an edge-set of an edge-set tree to form a new edge-set tree containing trees with the same length.

External net: A net with terminals in both two sets of partitioned terminals.

Fake Steiner point: A "Steiner point" that connects two nodes or Steiner points in an edge-set tree.

Figure of merit: A reference value for measuring a characteristic generated in placement or routing.

Fixed edge-set: An edge-set connecting two fixed vertices.

Force-directed placement: An iterative placement technique in which the routing length is minimized by simulating the lower energy state in a system of fictitious forces, proportional to distances of modules.

Forward search: A search process for providing progressing information which starts from one or more terminals and progresses with a specific search technique, until an object is encountered.

Graph: A super-set of a set of edges and a set of vertices used to describe the relationship between the edges and vertices.

Grid expansion: A search process that progresses from a grid to its adjacent grids in a gridded workspace.

Gridded workspace: A workspace divided into square grids for searching a path by a grid expansion technique.

Gridless searching: A routing technique that searches a path from the coordinates of obstacles, rather than the grids in a gridded workspace.

Gridless workspace: A workspace divided into obstacles that accomodate routed paths, and available areas.

Hadlock router: A maze-searching technique based on minimizing the total number of exhausted grids around obstacles for connecting two terminals.

Half-perimeter method: A technique that computes the routing length in placement by a half-perimeter of the minimum rectangle that encloses all terminals of a net.

Hightower's router: A line-searching technique that progresses from a fast line to another fast line, which is the longest line segment nearest the previous parallel fast line.

Initial placement: The placement configuration that is obtained by placing all un-placed modules into a workspace, and provided for the successive iterative placement process.

Interconnect connectivity: The wiring length per unit area required by a device for connecting with other devices.

Iso-distance error graph (IDEG): A gridded graph in which each location is associated with a connection error corresponding to a specific tree type and locations with the same connection error are connected by a trace.

Iterative placement: A placement technique that iteratively moves modules to improve the placement configuration and to approach given objectives.

Layer: A conduction sheet etched to form wires or a ground place in a printed wiring substrate.

Layout: Arranging modules and interconnecting electronic nets on a substrate.

Lee's (Lee-Moore's) router: A maze-searching technique for searching a shortest path, by simultaneously expanding grids to their adjacent grids until the source grid is encountered.

Line searching: A search technique that searches a path connecting two terminals from a set of line segments to other set of line segments, which are extended from a terminal, or imaginary grids occupied by an existing line segments.

Manhattan distance: A distance between two points, formally called a rectilinear distance, measured by a half-perimeter of the rectangle with the points as vertices.

Matched pair: A pair of edges with the same shape and the same length.

Matched-pair problem: Finding a pair of shortest paths with the same shape and the same length in a routing workspace.

Maze searching: A strategy for searching a path connecting two terminals in a gridded workspace.

Mikami-Tabuchi's router: A line-searching technique implemented by extending points in a set of line segments to another set of line segments in routing.

Min-cut placement: A placement technique where the cost function is the number of "cuts" generated by partitioning modules into sub-sets.

Minimum detour-length router: A routing technique based on minimizing the total detour length exhausted for routing around obstacles.

Minimum odd cycle cover: An odd cycle cover of minimum cardinality.

Minimum rectangle: The rectangle with the minimum perimeter enclosing a set of points.

Moving edge-set: An edge-set with at least one dynamic Steiner point as a vertex.

Multiplication rule: The rule of computing the number of ways generating event E_1 and event E_2.

Net: A physical tree in which an edge is a wire, and a vertex is a terminal or a module.

Net size (signal set): The number of terminals in a net.

NP-complete problem: An NP problem that can be transformed from all NP problems in polynomial time.

NP-hard problem: A decision problem that can be transformed from all NP problems in polynomial time, and is usually an optimization problem corresponding to an NP problem, but need not be an NP complete problem.

NP problem: A decision problem that can be solved by a nondeterministic algorithm in polynomial time.

NR edge-set tree: A no-redundancy edge-set tree in which neither a redundant-set between two adjacent edge-sets nor a redundant length of an arbitrary edge-set can exist.

Obstacle: Areas or grids occupied by routed paths in a routing workspace.

One-dimensional Steiner point: A dynamic Steiner point that can be moved along the horizontal or vertical direction.

Odd cycle cover: A set of edges whose removal leaves a graph free of an odd cycle.

Pair-linking placement: A constructive placement technique based on the connectivity between placed modules and unplaced modules.

Partitioning: A process that divides a set of terminals (nodes) into sub-sets for exploring the relationship among different subsets.

Path: A connection that connects two terminals with a sequence of line segments (physical sub-edges) and their neighboring clearances.

Permitted connectivity: The signal wire length, allowed per unit area of the substrate.

Permutation: Taking r elements from n $(n \geq r)$ distinct elements in order.

Placement: A technique that provides a module configuration for routing by placing modules on a workspace to satisfy one or more objectives.

Point-set of an edge-set: A union of points in all edges contained in an edge-set, that is a set of points enclosed exactly by a rectangle with the same vertices as those of the edge-set.

P problem: A decision problem that can be solved by a deterministic algorithm in polynomial time.

Quadratic assignment placement: A technique to find a placement configuration by minimizing $\sum_{u,v} c_{uv} d_{p(u)p(v)}$, where u and v are modules, $p(u)$ $p(v)$ are all permutations of u and v, respectively, c_{uv} is the connectivity (cost) of u and v, and $d_{p(u)p(v)}$ is the distance between u and v.

Rectilinear distance: A distance measured by a half-perimeter of the rectangle with two vertices (nodes, or terminals) as end-vertices, often called a Manhattan distance in technical publications.

Rectilinear tree: A tree with every edge length measured by the rectilinear distance.

Redundant length: The maximum length redunction obtained by re-connecting two edge-set subtrees generated by deleting an edge-set from an edge-set tree.

Redundant set: the intersection of two point-sets of adjacent edge-sets, that contains a set of two or more points.

Routable substrate area: The surface area that can be used for conductors, including traces, vias, via-pads and device lead pads.

Routing efficiency: The ability of the router to utilize all the available wiring tracks.

Routing length: A wire length needed by routing, where specific wires are not considered.

Routing track: A narrow stripe on a workspace used to accommodate a path or part of a path.

Row-based tree: A tree in which nodes on a row are primarily connected.

Signal set: A set of signal terminals in a net.

Simulated annealing placement: A placement technique where the iterative processes simulated the thermal equilibrium states are used to approach the placement objective.

Source-sink tree: A tree in which a specific node, called a source, connects all other nodes, called sinks.

Spanning tree: A tree connecting nodes without Steiner points.

Staircase layout: A technique connecting trees with each edge having at most one or two bends.

Steiner point: An additional point for connecting a set of nodes into a tree.

Steiner's problem: A combinatorial-graph problem for finding a tree connecting a set of nodes with the minimum tree length.

Steiner tree: A tree connecting a set of nodes with additional vertices called Steiner points.

Terminal: A pin, a joint, or a mount on a module used to electronically connect with other modules.

Time complexity: The time needed by an algorithm expressed as a function of the size of a problem.

Traveling salesman problem: A combinatorial problem for finding a shortest cycle to connect a set of vertices (cities).

Tree: A connected graph that connects a set of nodes without cycle.

Two-dimensional Steiner point: A dynamic Steiner point that can be moved along both the horizontal and vertical directions.

Via: An electrical conductor connecting a net routed on different layers.

Wire density: The signal-wire length per unit area of the substrtate.

Wire segment: A physical line segment in a net.

Index

Milton Keynes UK
Ingram Content Group UK Ltd.
UKHW020018071024
449327UK00032B/2830